DEVELOPMENTS IN SOLVENT EXTRACTION

ELLIS HORWOOD SERIES IN ANALYTICAL CHEMISTRY
Series Editors: Dr R. A. CHALMERS and Dr MARY MASSON, University of Aberdeen
Consultant Editor: Prof. J. N. MILLER, Loughborough University of Technology

S. Alegret	Developments in Solvent Extraction
S. Allenmark	Chromatographic Enantioseparation — Methods and Applications
G.E. Baiulescu, P. Dumitrescu & P.Gh. Zugravescu	Sampling
H. Barańska, A. Łabudzińska & J. Terpiński	Laser Raman Spectrometry
G.I. Bekov & V.S. Letokhov	Laser Resonant Photoionization Spectroscopy for Trace Analysis
K. Beyermann	Organic Trace Analysis
O. Budevsky	Foundations of Chemical Analysis
J. Buffle	Complexation Reactions in Aquatic Systems: An Analytical Approach
D.T. Burns, A. Townshend & A.G. Catchpole	Inorganic Reaction Chemistry Volume 1: Systematic Chemical Separation
D.T. Burns, A. Townshend & A.H. Carter	Inorganic Reaction Chemistry: Volume 2: Reactions of the Elements and their Compounds: Part A: Alkali Metals to Nitrogen, Part B: Osmium to Zirconium
J. Churáček	New Trends in the Theory & Instrumentation of Selected Analytical Methods
E. Constantin, A. Schnell & M. Mruzek	Mass Spectrometry
R. Czoch & A. Francik	Instrumental Effects in Homodyne Electron Paramagnetic Resonance Spectrometers
T.E. Edmonds	Interfacing Analytical Instrumentation with Microcomputers
Z. Galus	Fundamentals of Electrochemical Analysis, Second Edition
S. Görög	Steroid Analysis in the Pharmaceutical Industry
T. S. Harrison	Handbook of Analytical Control of Iron and Steel Production
J.P. Hart	Electroanalysis of Biologically Important Compounds
T.F. Hartley	Computerized Quality Control: Programs for the Analytical Laboratory
Saad S.M. Hassan	Organic Analysis using Atomic Absorption Spectrometry
M.H. Ho	Analytical Methods in Forensic Chemistry
Z. Holzbecher, L. Diviš, M. Král, L. Šůcha & F. Vláčil	Handbook of Organic Reagents in Inorganic Chemistry
A. Hulanicki	Reactions of Acids and Bases in Analytical Chemistry
David Huskins	Electrical and Magnetic Methods in On-line Process Analysis
David Huskins	Optical Methods in On-line Process Analysis
David Huskins	Quality Measuring Instruments in On-line Process Analysis
J. Inczédy	Analytical Applications of Complex Equilibria
M. Kaljurand & E. Küllik	Computerized Multiple Input Chromatography
S. Kotrlý & L. Šůcha	Handbook of Chemical Equilibria in Analytical Chemistry
J. Kragten	Atlas of Metal-ligand Equilibria in Aqueous Solution
A.M. Krstulović	Quantitative Analysis of Catecholamines and Related Compounds
F.J. Krug & E.A.G. Zagotto	Flow Injection Analysis in Agriculture & Environmental Science
V. Linek, V. Vacek, J. Sinkule & P. Beneš	Measurement of Oxygen by Membrane-Covered Probes
C. Liteanu, E. Hopîrtean & R. A. Chalmers	Titrimetric Analytical Chemistry
C. Liteanu & I. Rîcă	Statistical Theory and Methodology of Trace Analysis
Z. Marczenko	Separation and Spectrophotometric Determination of Elements
M. Meloun, J. Havel & E. Högfeldt	Computation of Solution Equilibria
M. Meloun, J. Militky & M. Forina	Chemometrics in Instrumental Analysis: Solved Problems for IBM PC
O. Mikeš	Laboratory Handbook of Chromatographic and Allied Methods
J.C. Miller & J.N. Miller	Statistics for Analytical Chemistry, Second Edition
J.N. Miller	Fluorescence Spectroscopy
J.N. Miller	Modern Analytical Chemistry
J. Minczewski, J. Chwastowska & R. Dybczyński	Separation and Preconcentration Methods in Inorganic Trace Analysis
T.T. Orlovsky	Chromatographic Adsorption Analysis
D. Pérez-Bendito & M. Silva	Kinetic Methods in Analytical Chemistry
B. Ravindranath	Principles and Practice of Chromatography
V. Sedivec & J. Flek	Handbook of Analysis of Organic Solvents
O. Shpigun & Yu. A. Zolotov	Ion Chromatography in Water Analysis
R.M. Smith	Derivatization for High Pressure Liquid Chromatography
R.M. Smith	Handbook of Biopharmaceutic Analysis
K.R. Spurny	Physical and Chemical Characterization of Individual Airborne Particles
K. Štulík & V. Pacáková	Electroanalytical Measurements in Flowing Liquids
J. Tölgyessy & E.H. Klehr	Nuclear Environmental Chemical Analysis
J. Tölgyessy & M. Kyrš	Radioanalytical Chemistry, Volumes I & II
J. Urbanski, *et al.*	Handbook of Analysis of Synthetic Polymers and Plastics
M. Valcárcel & M.D. Luque de Castro	Flow-Injection Analysis: Principles and Applications
C. Vandecasteele	Activation Analysis with Charged Particles
F. Vydra, K. Štulík & E. Juláková	Electrochemical Stripping Analysis
N. G. West	Practical Environmental Analysis using X-ray Fluorescence Spectrometry
J. Zupan	Computer-supported Spectroscopic Databases
J. Zýka	Instrumentation in Analytical Chemistry

DEVELOPMENTS IN SOLVENT EXTRACTION

Editor:
S. ALEGRET
Department of Analytical Chemistry
Universitat Autónoma de Barcelona, Spain

Translation Editor:
MARY R. MASSON
Department of Chemistry, University of Aberdeen

Published for the
INSTITUT D'ESTUDIS CATALANS
Barcelona
by

ELLIS HORWOOD LIMITED
Publishers · Chichester

First published in 1988 by

CHEMISTRY

ELLIS HORWOOD LIMITED
Market Cross House, Cooper Street,
Chichester, West Sussex, PO19 1EB, England
The publisher's colophon is reproduced from James Gillison's drawing of the ancient Market Cross, Chichester.

Distributors:

Australia and New Zealand:
JACARANDA WILEY LIMITED
GPO Box 859, Brisbane, Queensland 4001, Australia

Canada:
JOHN WILEY & SONS CANADA LIMITED
22 Worcester Road, Rexdale, Ontario, Canada

Europe and Africa:
JOHN WILEY & SONS LIMITED
Baffins Lane, Chichester, West Sussex, England

North and South America and the rest of the world:
Halsted Press: a division of
JOHN WILEY & SONS
605 Third Avenue, New York, NY 10158, USA

South-East Asia
JOHN WILEY & SONS (SEA) PTE LIMITED
37 Jalan Pemimpin # 05–04
Block B, Union Industrial Building, Singapore 2057

Indian Subcontinent
WILEY EASTERN LIMITED
4835/24 Ansari Road
Daryaganj, New Delhi 110002, India

© **1988 S. Alegret/Ellis Horwood Limited**

British Library Cataloguing in Publication Data
Developments in solvent extraction.
1. Solvent extraction.
I. Alegret, S. Salvador.
II. Institut d'Estudis Catalans
541.3′482

Library of Congress CIP also available

ISBN 0–7458–0303–2 (Ellis Horwood Limited)
ISBN 0–470–21251–9 (Halsted Press)

Typeset in Times by Ellis Horwood Limited
Printed in Great Britain by Hartnolls, Bodmin

Table of contents

1

Solvent extraction in analytical chemistry and separation science

Henry Freiser
Department of Chemistry, University of Arizona, Tucson, Arizona 85721

Solvent extraction has long enjoyed a well deserved position of prominence among analytical chemists as a powerful separation technique applicable both to trace and macro levels of materials. Solvent extraction chemistry has developed in a most dramatic way in the last quarter of a century. Work in this area has not only provided the basis for a rich store of analytical methodology characterized by high sensitivity and selectivity, but has also illuminated many fundamental aspects of a wide variety of inorganic co-ordination complex reactions [1–4]. Because of the great range of concentrations from 'weightless' trace levels of carrier-free radio-isotopes to the macro levels of several weight per cent of metal ions in which quantitative separations by solvent extraction is applicable, this technique is equally useful in both analytical and preparative, e.g. process scale , modes. The continuing vigour of this interesting field is attested by the continuing high publication rate of research reports from active groups throughout the world (about 1000 papers annually) [5–6]. It is not daring to predict that much innovative and fruitful research in both fundamental and applied research will be conducted in the next decade or two.

Despite the great variety of extraction systems it is possible to describe every extraction by a simple, three step scheme (Table 1.1).

Table 1.1 — Process of Extraction

1. Chemical interactions in the aqueous phase
2. Phase distributions of extractable species
3. Chemical interactions in the organic phase

Metal salts are generally soluble in aqueous media not only because of the high dielectric constant of water, which readily permits dissociation of ionic species, but also, more importantly, because the basic nature of water results in the solvation of metal ions. Hydration of metal ions provides them with a 'protective' solvent sheath. The role of the extractant is to form a metal complex by supplanting the water to give a species that is more likely to be compatible with organic solvents. This is so important that classification of metal extractions systems is based on the nature of the complex involved (Table 1.2). In this report, attention will be focussed on chelating extractants (see 3, 3–4, and 1–3 in Table 1.2).

Table 1.2 — Types of extractable complexes

Co-ordination complexes	1. Simple (monodentate) e.g., $GeCl_4$
	2. Heteropoly acid, e.g., $H_3PO_4.12\,MoO_3$
	3. Chelate (polydentate) e.g., Fe (oxinate)$_3$
Ion pair complexes	4. Simple ion pair, e.g., Cs^+, $(C_6H_5)_4B^-$
Mixed types	1.4 $(Onium)^+$, $FeCl_4^-$ or MnO_4^-
	3.4 $Cu(Neocup)^{2+}$, ClO_4^- or $3(R_4N)^+$, $Mg(Oxinate)_3^-$
	1.3 $Th(TTA)_4.TBP$

Extraction Equilibria. Solvent extraction equilibria involved in metal chelate systems can be described by means of Fig. 1.1, which is useful as a frame of reference in a discussion of areas of current research on extraction.

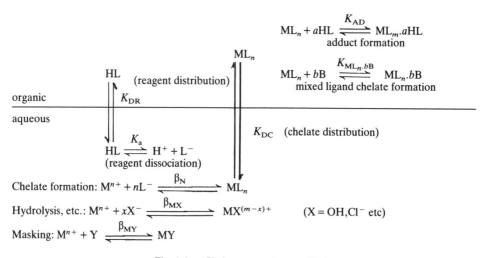

Fig. 1.1 — Chelate extraction equilibria.

The distribution ratio, D_M, is affected by all of these component reactions which are taken into account in the expression describing the overall extraction equilibrium

$$D_M \equiv \frac{C_M(o)}{C_M} = \beta_N \frac{K_a^n K_{DC}}{K_{DR}^n} \frac{[HL]_o^n}{[H+]^n} \cdot \left\{ 1 + K_{AD}[HL]_o^a + K_{ML}[B]_o^b \right\} \alpha_M$$

$$(1.1)$$

where α_M is the fraction of metal present as the hydrated metal ion, M^{n+} in the aqueous phase and, as seen from equation (2), can readily be evaluated in terms of the various βs and reagent concentrations:

$$\alpha_M = \left[1 + \sum_{i}^{N} \beta_i [L]^i + \sum_{i}^{x} \beta_{MX_i}[X]^i + \beta_{MY}[Y] \right]^{-1} \tag{1.2}$$

Figure 1.1 demonstrates that metal separation processes that utilize solvent extraction provide many opportunities for 'fine tuning' by means of appropriate adjustment of chemical parameters. It can be seen at a glance that chemical interactions in both the aqueous and organic phases dominate extraction equilibrium and, therefore, that study of such reactions (Fig. 1.2) can lead to improved separation efficiency. Thus, it is possible to exploit not only inherent differences in the ability of metal ions to form metal chelates arising from changes in Lewis acidity associated with their position in the periodic table and varying with their oxidation state, but also to modify such ability by varying the nature of the bonding atoms [7–9] of the chelating agent, the chelate ring size and other steric factors, all reflected in values of β and K_a. The effects of replacing oxygen by sulphur or selenium are particularly significant.

$$M(H_2O)_n^{2+} + 2 \; \overset{B_1}{\underset{B_2}{\diagdown}}(X)_y \; \overset{\beta_2}{=} \; (X)_y \overset{B_1}{\underset{B_2}{\diagup}} M \overset{B_2}{\underset{B_2}{\diagdown}}(X)_y + nH_2O$$

$$B = O, S, N, P \quad X = C \ldots y = 2,3$$

Sulphur analogues of oxygen-containing chelating agents are generally more selective (increased stability) for the 'softer' metal ions, less selective for the 'harder' ions), and have lower proton affinities, resulting in higher proton displacement constants ($K_{PD} = \beta_n K_a{}^n$). An interesting example can be seen in the analogues of 8-quinolinol. As seen in Fig. 1.3, the increase in $\log \beta$ from Mn to Zn is significantly higher for 8-mercaptoquinoline [10]. This trend is even more pronounced with the selenium analogue [9]. In the same series, the role of steric hindrance in reducing β values is illustrated by comparing the effects produced by a methyl group (Me) in the 2-position (Table 1.3). Despite the larger radius of the S atom, stability decrease in the 2-Me analogue resembles that in the 2-methyl-8- quinolinol [10]. Zinc is not so affected because of the tetrahedral configuration.

Further differentiation can be accomplished for metals exhibiting variable co-ordination numbers, permitting the formation of self-adduct or mixed-ligand complexes (note role of K_{AD} and $K_{ML_n}.bB$). Masking agents can also significantly assist in improving selectivity. Distribution parameters K_{DR} and K_{DC} play significant roles

Characteristic Of Reagent
  1. RING SIZE
  2. REAGENT BASICITY
  3. ELECTRONEGATIVITY OF DONOR ATOMS (HARD/SOFT)
  4. NUMBER OF RINGS FORMED (CHELATE EFFECT)

Characteristic Of Metal
  1. ELECTROSTATIC BONDING   3. LIGAND FIELD EFFECTS
     Acidity related to Z^2/R
  2. COVALENT BONDING
     Acidity related to
     Electronegativity (HARD/SOFT)
     Position of available orbitals

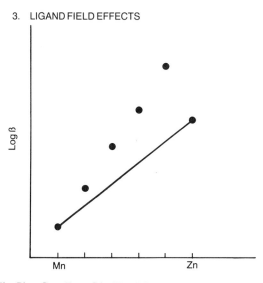

$$Pd > Cu > Ni > Pb > Co > Zn > Cd > Fe > Mn > Mg$$

Characteristic of Chelate
  1. RESONANCE EFFECT
  2. STERIC FACTORS

Fig. 1.2 — Factors affecting chelate stability

in the chelate extraction process. For example, increasing values of (K_{DC}/K_{DR}^n) will increase the extraction constant. Generally, structural changes in the ligand to increase K_{DC} for the chelate will also serve to increase K_{DR} for the ligand. One interesting potential research area is to find ligands which undergo reaction in the aqueous phase (other than proton transfer, of course) to which the chelate is not subject, e.g. some tautomeric equilibrium (neutral ⇌ zwitterion). This would serve to reduce the value of K_{DR} without affecting K_{DC}, thus enhancing the extraction constant as well as the limiting sensitivity given by K_{DC}. 8-Mercaptoquinoline and its analogues, which are zwitterionic in water but neutral in organic solvents, exhibit this useful effect (Fig. 1.4 and Table 1.4) [10] but this represents a relatively unexplored and fruitful area. Thanks largely to the work of Hansch [11], it is now possible to make reasonable estimates of values for K_{DR} for various compounds from examin-

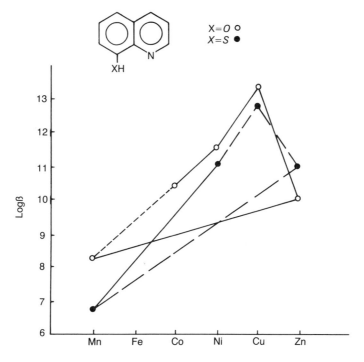

Fig. 1.3 — Role of electronegativity of bonding atom

Table 1.3 — Stability constants of 1:1 metal oxinates and thio-oxinates

Metal Ion	QSH	2MeQSH	QOH	2MeQOH
Co(II)	7.9	9.6	10.55	9.63
Ni(II)	11.0	9.2	11.44	9.41
Cu(II)	22.7	11.7	13.49	12.48
Zn(II)	11.0	11.1	9.96	9.82

ation of their structural elements. It is possible to predict, to within ± 0.1 log unit, K_{DR} values for 8-quinolinol and its various substituted derivatives, and almost as well for other extractant families.

Of the chemical reactions involved in chelate extraction, naturally that of chelate formation is most central. Much systematic work has resulting in fairly well-defined

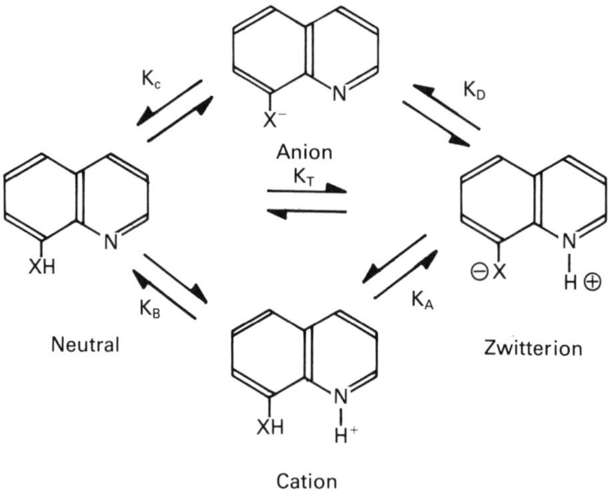

Fig. 1.4 — Chemical forms of 8-mercaptoquinolines

Table 1.4 — Dissociation constants of substituted 8-mercaptoquinolines in aqueous solution at 25°C, $\mu = 0.10^*$

substituent	pK_A	pK_B	pK_C	pK_D	$\log K_T$
parent	2.07	3.50	6.84	8.27	1.43
2-CH$_3$	2.29	4.08	7.26	9.05	1.79
4-CH$_3$	2.07	4.27	6.97	9.18	2.20
6-CH$_3$	2.12	3.86	6.74	8.88	1.74
7-CH$_3$	2.25	5.57	5.93	9.27	3.32
2-4-di-CH$_3$	2.24	4.88	7.41	10.05	2.64
4-6-di-CH$_3$	2.30	4.54	7.06	9.30	2.25
2-C$_6$H$_5$	1.37	2.11	6.98	7.81	0.85

* Kawase (1966)

principles (Fig. 1.2) that describe the effect of structural changes in the ligand and of the nature of the metal ion on β_N values [14]. On the other hand, very little work has been done on the systematic scrutiny of the structural effects on selectivity, i.e. on the ratio of β_N values of given pairs of metal ions. The combination of a suitable pair of reagents, a particular extractant with a particular masking agent, may serve to enhance the separation of a given metal–ion pair well beyond that achievable with either reagent alone. Examination of the behaviour of extractants having (a) some adverse steric factor, or those providing (b) special d–π bonding possibilities, is an active and fruitful area of research. Our extraction research has addressed these points as well as the examination of selectivity that can be achieved through the use of systems in which adduct or mixed-ligand chelate formation (characterized by K_{AD} and K_{ML}) can occur [12–16]. That is, in addition to the great differences that are attainable when one metal extracts as a simple chelate, ML_n and the other as an adduct, separation can also be improved when both extracted species are adducts, provided that appropriately different K_{AD} values obtain. We are in the course of a systematic investigation of the separation of individual lanthanide ions, which is one of the most challenging of inorganic separation problems, in which use is made of this strategy. Lanthanides have high co-ordination numbers, so they are likely to form adducts or mixed ligand chelates. In our recent work, we have studied these systems by looking at their electrochemical behaviour [17–21]. We intend to continue along similar lines and further improvements in selectivity may be expected.

REFERENCES

[1] G. H. Morrison and H. Freiser, *Solvent Extraction in Analytical Chemistry*, Wiley, New York, 1957.
[2] J. Stary, *Solvent Extraction of Metal Chelates*, Pergamon Press, Oxford, 1964.
[3] Y. A. Zolotov, *Extraction of Metal Chelates*, Ann Arbor, London, 1970.
[4] Y. Marcus and A. S. Kertes, *Ion Exchange and Solvent Extraction of Metal Compounds*, Wiley, New York, 1969.
[5] H. Freiser, *Anal. Chem.*, 1968, **40**, 522R.
[6] T. Sekine and S. Hasegawa, *Solvent Extraction Chemistry*, Dekker, New York, 1977.
[7] E. Sekido, Q. Fernando and H. Freiser, *Anal. Chem.*, 1963, **35**, 1550.
[8] E. Sekido, Q. Fernando and H. Freiser, *Anal. Chem.*, 1964, **36**, 1768.
[9] E. Sekido, Q. Fernando and H. Freiser, *Anal. Chem.*, 1965, **37**, 1556.
[10] A. Kawase and H. Freiser, *Anal. Chem.*, 1967, **39**, 22.
[11] C. Hansch and A. J. Leo, *Substituent Constants for Correlation Analysis in Chemistry and Biology*, Wiley, New York, 1979.
[12] T. Hori, M. Kawashima and H. Freiser, *Separ. Sci. Tech.*, 1980, **15**, 861.
[13] M. Kawashima and H. Freiser, *Anal Chem.*, 1981, **53**, 284.
[14] O. Tochiyama and H. Freiser, *Anal Chem.*, 1981, **53**, 874.
[15] E. Yamada and H. Freiser, *Anal Chem.*, 1981, **53**, 2115.
[16] O. Tochiyama and H. Freiser, *Anal. Chim. Acta*, 1981, **131**, 233.
[17] Y. Sasaki and H. Freiser, *Inorg. Chem.*, 1983, **22**, 2289.
[18] S. Lin and H. Freiser, *Anal. Chem.*, 1987, **59**, 2834.
[19] S. Umetani and H. Freiser, *Inorg. Chem.*, 1987, **26**, 3179.
[20] S. Taguchi, M. Hojjatie and H. Freiser, *Anal. Chim. Acta*, 1987, **197**, 333.
[21] W. Yu and H. Freiser, *Talanta*, in the press.

2

The thermodynamics of solvent extraction

Y. Marcus
Department of Chemistry, The Hebrew University of Jerusalem, Jerusalem 91904. Israel

2.1 BASIC THERMODYNAMIC RELATIONSHIPS RELEVANT TO SOLVENT EXTRACTION

Distribution equilibria are described by the distribution quotient, which is a function not only of the concentration of the distribuend and of substances that react with it, but also of other substances in the solution, and of external constraints such as the temperature. The distribution equilibrium constant is a quantity descriptive of the distribution equilibrium that is independent of the concentrations of the distribuend and the other reactants. Since it does depend on the other conditions, i.e., the concentrations of other constituents of the solution, it is designated as a "conditional" equilibrium constant. If account is taken also of these dependencies, then a 'true' or 'thermodynamic' equilibrium constant is obtained, that depends only on the temperature and on the chemical nature of the solvents and the distribuend. Thermodynamics is the methodology through which these dependencies are unraveled, and the fundamental chemistry of the distribution reaction can be studied from the quantities that are measurable experimentally.

Thermodynamics deals with components (not species), i.e., with substances that can be added independently to the system. It is impossible, for instance, to add UO_2^{2+} ions to a system, one can only add a salt, such as $UO_2(NO_3)_2 \cdot 6H_2O$. It is impossible to add the dimer $(CH_3CO_2H)_2$ to a system, one can only add the substance 'acetic acid'. It is impossible to reduce $Fe^{3+}(aq)$ in a system (i.e., add electrons to it), without oxidizing some other substance in it. The concept of 'component' in the thermodynamic sense is essential for the understanding of the phase behaviour of distribution systems.

A phase, in the thermodynamic sense, is a region of space that has uniform properties throughout. Distribution operations are generally conducted at atmospheric pressure, and the gas phase is irrelevant to them. In this case the system involves two phases and at least three components: the distribuend and two mutually immiscible liquid solvents. The phase rule of Gibbs tells us what the variance of such a system is, i.e., how many variables can be chosen at will in order for the equilibrium state of the system to be defined. The general form of this rule is $V = 2 + C - P$, where V is the variance, C is the number of components, and P the number of phases. For our purpose, since the gas phase and the external pressure are irrelevant and $P = 2$, we have $V = C - 1$. We can, for example, fix the temperature and the concentration of the single distribuend in one of the liquid phases arbitrarily. Since for *this* system $V = 3 - 1 = 2$, we have thereby exhausted the quantities that can be chosen at will, so the concentration of the distribuend in the second liquid phase is fixed by the equilibrium condition, and cannot be chosen at will [1]. There occur rare cases in liquid–liquid distribution where three liquid phases coexist. If there are only three components (i.e., a single solute and two solvents) the variance is then $V = 1$, and at a given temperature the composition of all three liquid phases is fixed. Variable compositions require the presence of at least a fourth component.

In liquid systems it is convenient to distinguish between 'mixtures', where all components are liquid, and 'solutions', where one component, the 'solvent', is liquid and the others are solids (or, rarely, gases), and are called 'solutes'. In distribution systems attention is generally focused on one solute, the 'distribuend', that distributes between two mutually immiscible liquids. Other solutes in one or both phases may react with the distribuend (reactants, ligands, adduct-formers, etc.) or be relatively inert, i.e., not enter into the formation of new species with the distribuend. Such an inert substance, if liquid, may act as a 'diluent' of the solvent, the latter acting as the 'extractant'. The solution of the extractant (generally a liquid, but occasionally a solid) in the diluent is then the solvent.

The composition of a liquid phase is expressed in terms of mass fractions $w_i = m_i'/\Sigma_j m_j'$, mole fractions $x_i = (m_i'/M_i)/\Sigma_j(m_j'/M_j) = n_i/\Sigma_j n_j$, or volume fractions $\phi_i = V_i/\Sigma_j V_j = (m_i'/d_i)/\Sigma(m_j'/d_j)$, where m' is the mass, M the relative molar mass, n the amount of substance (number of moles), V the volume, and d the density of the designated component. The composition of a solution can also be described in terms of the molality $m_i = n_i/m'_{\text{solvent}}$, where the mass of the solvent is expressed in kg, or the molarity $c_i = n_i/V_{\text{solution}}$, where the volume is expressed in litres. The 'nominal' composition is given, on any of the above listed composition scales, by a specification in terms of the components [2]. It must be realized that chemical reactions among the components produce constraints. One such constraint, already mentioned, is the electroneutrality requirement of the system and of each phase. The chemical reactions can, within these constraints, produce various species. The specification of the concentrations of the species that exist in the solution provides its 'actual' composition. Thus the dissociation of an electrolyte (that is a component) into its constituent ions produces new species that may react further, e.g., with the solvent (water) by hydrolysis, etc., to produce a final set of species that are in equilibrium [2].

The behaviour of a solution, i.e. the change in its properties with the change in its composition or the temperature, is conveniently referred to a hypothetical state — the 'ideal solution' — and to the deviations from it — its 'excess functions'. Only the former will be discussed in this section, a discussion of the latter being deferred to section 2.2. An ideal mixture (solution) is defined as one where the mixing does not change the enthalpy H and the volume V, which are the sum of the products of these molar quantities of the components with their numbers of moles: $H = \Sigma_j n_j H_j^*$, $V = \Sigma_j n_j V_j^*$ or $\Delta H^{\text{M}} = 0$ and $\Delta V^{\text{M}} = 0$. Here an asterisk designates a pure component and superscript ^{M} the change on mixing. A further characterization of the ideal mixture is that the entropy of mixing is $\Delta S^M = -R\Sigma_j x_j \cdot \ln x_j$ where R is the gas constant. The Gibbs free energy of the mixture is therefore $G = \Sigma_j n_j G_j^* + RT\Sigma_j x_j \ln x_j$ [2].

An ideal dilute solution conforms to Raoult's law: the vapour pressure of the solvent is given by $p_{solvent} = p^*_{solvent} (1 - \Sigma_j x_{solutes\, j})$. It also conforms to Henry's law, which says that the vapour pressure of each solute is proportional to its mole fraction: $p_{solute\, i} = K_i x_i$, where K_i is the Henry's law constant; this generally does *not* equal the vapour pressure p_i^* of the pure solute [2].

The distribution of a distribuend between two liquid phases can be described in terms of its distribution isotherm: its mole fraction in the one phase, $\bar{x}_i$, plotted against that in the other, x_i, at a given temperature. (If one of the phases is aqueous, then it is customary to designate the other, generally organic, by a bar above the symbol.) It is an observed fact that such plots tend to start out from the origin as a straight line. The same holds also for compositions expressed on other scales, e.g., the molarity one. The ratio of the concentrations in the two phases is designated as the 'distribution ratio': $D_{i(x)} = \bar{x}_i/x_i$ or $D_{i(c)} = \bar{c}_i/c_i$. The observed initial linearity of the distribution isotherm is an expression of Nernst's distribution law:

$$\lim_{x\, or\, c \to 0} D_i = P_i,$$

where P_i is the partition constant of the distribuend i. Some systems that conform to this ideal behaviour are $HgBr_2$ distributing between toluene and water, RuO_4 distributing between methyl isobutyl ketone and water, and thorium tetrakis-(thenoyltrifluoroacetone) between chloroform and water.

What quantity describes the tendency for a solute to transfer from one liquid phase to the other? In other words, what is the driving force for mass transfer in a distribution system? The relevant quantity is the 'chemical potential' of the distribuend, and as long as this is larger in one phase than in the other, the distribuend will transfer form the former to the latter. The transfer causes changes in the chemical potentials in the direction of their equalization, and the transfer will cease when the chemical potentials are the same in the two phases. This is the condition for distribution equilibrium. Mathematically expressed, the chemical potential is $\mu_i = (\partial G_{mixture}/\partial n_i)_{T,P,n_{j\neq i}}$. Thermodynamics teaches that the chemical potentials of components do not change independently as the composition is varied. The Gibbs–Duhem law says that at constant temperature (and pressure) $\Sigma n_j d\mu_j = 0$. This relationship permits the evaluation of the chemical potential of the solute from that of the solvent in a binary solution, a fact that is employed, for instance, in the osmometric method for the determination of the nominal activity coefficient and aggregation of a solute (see section 2.2).

The condition for distribution equilibrium can now be expressed as $\mu_{iI} = \mu_{iII}$, where I and II denote the two phases. It remains to be seen how μ_i depends on the concentration of i and on the temperature.

It is convenient to define a 'standard chemical potential' as the limit $\mu_{i(x)}^\circ = \lim_{x \to 0} (\mu_i - RT \ln x_i)$, for the rational (mole fraction) scale, or as $\mu_{i(c)}^\circ = \lim_{c \to 0} (\mu_i - RT \ln c_i)$, for the molar scale. It is seen that the standard chemical potential depends on the concentration scale employed. Ideal solutions can be defined now as those for which the expression $\mu_i = \mu_{i(c)}^\circ + RT \ln c_i$ (or similarly for the x-scale) is valid. The distribution law can now be expressed in terms of the standard chemical potentials: from $\mu_{iI} = \mu_{iI}^\circ + RT \ln c_{iI} = \mu_{iII} = \mu_{iII}^\circ + RT \ln c_{iII}$ one obtains $D_i = c_{iI}/c_{iII} = \exp[(\mu_{iII}^\circ - \mu_{iI}^\circ)/RT] = P_i$. The partition constant is thus exp of $(1/RT)$ times the difference between the standard chemical potentials of the distribuend in the two phases.

The temperature dependence of the partition constant in an ideal distribution system determines the standard entropy and enthalpy changes for the distribution equilibrium. At constant pressure $\Delta S_{i,I\,II}^\circ = -d(\mu_{iII}^\circ - \mu_{iI}^\circ)/dT = -R \ln P_i + RT(d \ln P_i/dT)$ and $\Delta H_{i,I\,II}^\circ = (\mu_{iII}^\circ - \mu_{iI}^\circ) + T\Delta S_{i,I\,II}^\circ = -RT^2(d \ln P_i/dT)$. The standard enthalpy change can also be obtained calorimetrically or from the difference between the heats of solution of the distribuend in two separate phases (the two mutually saturated immiscible liquids, or the two pure liquids if their solubility is negligible), provided the system behaves ideally.

Even in systems behaving ideally in terms of the actual species, the behaviour in terms of the nominal components may appear non-ideal. This is the case if equilibria take place among the species, e.g., association of a species with itself to form oligomers, or with the solvent to form solvates, or with another component (species) to form adducts. The 'stoichiometric' distribution ratio is what is generally measured experimentally: $D = (\Sigma$ all distribuend species in phase I$)/(\Sigma$ all distribuend species in phase II). This is also often called the 'analytical' distribution ratio. The equilibria that the species undergo affect the observed value of D.

An example is the protonation of a basic distribuend or the deprotonation of an acidic one (or both of an amphoteric one), that make D depend on the pH. If these reactions are confined to, say, phase II (the aqueous phase), and if, say, only the neutral-base species can distribute but not the protonated one, then as the pH increases and protonation decreases D will rise (log D rises linearly with pH) until eventually only the neutral base exists at high pH and D levels off to a constant value [3].

Another example is the self-aggregation of the solute in the organic phase. This will cause D to increase with rising concentrations of the distribuend. If only a dimer is formed then the rate of the change of D_i with $\bar{c}_i$ is a constant: $d^2D_i/d\bar{c}_i^2 = 0$, but if higher aggregates are also formed then this second derivative is positive. The mean aggregation number of the distribuend in the organic phase can be calculated from the distribution curve, D_i vs. $\bar{c}_i$ [3].

These reactions that species of the distribuend may undergo in the solution have their own standard enthalpy and entropy changes. Therefore $-RT^2(d \ln D_i/dT)$ is, in general, not a measure of the enthalpy change of the partition equilibrium itself, and includes the enthalpic effects of these side reactions [4]. In order to separate the various effects, it is advisable to define a conditional extraction constant, K_{ex}', that is the (ideal) equilibrium constant for the extraction reaction of interest. From this are derivable the standard (conditional, ideal) Gibbs free energy of the extraction reaction $\Delta G_{ex}' = -RT \ln K_{ex}'$ and its standard (conditional, ideal) enthalpy change, $\Delta H_{ex}' = -RT^2(d \ln K_{ex}'/dT)$.

Ideal solutions are an idealization and cannot, as a rule, be realized in practice.

Nevertheless, there are systems that in dilute solutions approach the behaviour of ideal solutions. These may be treated according to the concepts developed in this lecture. For the majority of practical distribution systems this can be done only on extrapolation to infinite dilution. If this extrapolation can be done reliably (which is doubtful in the general case), then this limit of infinite dilution can be treated as discussed above.

2.2 THE EFFECTS OF NON-IDEALITIES IN THE AQUEOUS AND ORGANIC PHASES

Ideal solutions are characterized by the absence of interactions among the solute species that are not expressible in terms of definite, stoichiometric chemical reactions. Such interactions, generally classified as 'physical', strongly affect the behaviour of the system and cannot as a rule be ignored. They are expressed by means of excess functions. Thus, the chemical potential of a component in a non-ideal solution is given by $\mu_i = \mu_{i(c)}^{\circ} + RT \ln c + \mu_{i(c)}^{E}$, the last term being the excess of chemical potential (on the molar scale). The latter, in turn, can be expressed in terms of an activity coefficient: $y_i = \exp(\mu_{i(c)}^{E}/RT)$. Other concentration scales have their corresponding activity coefficients, e.g. $\gamma_i = (\mu_{i(m)}^{E}/RT)$ for the molality scale and $f_i = \exp(\mu_{i(x)}^{E}/RT)$ for the rational one. The product of the concentration and the corresponding activity coefficient is the activity of the component: $a = cy$. In aqueous solutions of electrolytes the activity of the solvent, the water, is generally expressed on the rational scale and approaches unity in dilute solutions, that of the electrolyte on the molal scale (this is generally tabulated in compilations) or, less often, the molar scale, and values for γ (or y) tend to unity as the concentration diminishes.

For distribution equilibrium in non-ideal solutions it is still true that $\mu_{iI} = \mu_{iII}$, but now $\exp[\mu_{iII}^{\circ} - \mu_{iI}^{\circ})/RT] = a_{iI}/a_{iII}$ is the ratio of the activities, not of the concentrations. The latter, i.e., the distribution ratio, is $D_i = c_{iI}/c_{iII} = \exp[\mu_{iII}^{\circ} - \mu_{iI}^{\circ})/RT](y_{iII}/y_{iI})$. Hence, it is necessary to evaluate the activity coefficients of the distribuend in the two phases in order to describe the distribution equilibrium thermodynamically.

It is expedient for the purpose of this section to limit the attention to distribution systems where one of the phases is aqueous and the other organic. It is also expedient to treat each phase separately, for the evaluation of the activity coefficients.

Since electrolytes dissociate in aqueous solutions into their constituent ions, it is the mean ionic activity coefficient $y_{\pm}$ that is relevant for them. This is given in moderately dilute solutions (up to a few tenths of moles per litre) by the Debye–Hückel–Davis expression: $\log y_{\pm} = -A|z_+ z_-|I^{\frac{1}{2}}(1 + B\mathring{a}I^{\frac{1}{2}})^{-1} + bI$. Here I is the ionic strength of the solution: $I = \frac{1}{2}\Sigma_i z_i^2 c_i$, where the summation extends over all ionic species present, cations and anions, z_i is the charge (in the algebraic sense) of the ion, A and B are constants independent of the electrolyte (they depend on temperature) and $\mathring{a}$ and b are electrolyte-dependent parameters [5,6]. The product $B\mathring{a}$ is sometimes taken as a universal constant($\sim 1.5 \, l^{\frac{1}{2}} \, mole^{-\frac{1}{2}}$), and then b is left the task of fitting the experimental data. According to the Debye–Hückel–Davis expression $\log y_{\pm}$ decreases with increasing electrolyte concentration, but not as sharply as it would if the term linear in I were absent.

For more concentrated aqueous electrolyte solutions, ion hydration ought to be taken into account explicitly. The above relationship of the activity coefficient is now better expressed on the molal scale (i.e., $\gamma_\pm$ and $I = \frac{1}{2}\Sigma_i z_i^2 m_i$), and the term linear in ionic strength is now formulated as $0.0078(h-v)m$, where the numerical coefficient is the molar mass of the solvent water (in kg) divided by $\ln 10$, h is the hydration number of the electrolyte, and v its stoichiometric coefficient ($v = 2$ for a 1:1 electrolyte, such as HCl, $v = 3$ for a 2:1 electrolyte, such as $MgCl_2$, etc.). The hydration numbers of electrolytes should, ideally, be the sums of those of the constituent ions, but in practice are only approximately so. They increase with the charge of the ion and as its size diminishes. The effect of the term linear in molality of the electrolyte is to cause the $\log \gamma_\pm = f(m)$ curve to bend up again after it reaches a minimum at 1–3m. In very concentrated solutions of highly hydrated electrolytes $\gamma_\pm$ is apt to reach very high values, of the order of hundreds or a few thousands [5].

The activity coefficient of an electrolyte in a mixture of electrolytes is dependent on the concentrations of all the others. This dependence can generally be described in terms of Harned's rule: $\log y_{\pm i} = \log y_{\pm i}^\circ + \Sigma_j \alpha_{ij} x_{j(I)}$. Here $y_{\pm i}^\circ$ is the activity coefficient of the electrolyte i if it were present at the ionic strength of the actual mixture, $x_{j(I)}$ is the fractional contribution of electrolyte $j \neq i$ to the ionic strength of the mixture, and α_{ij} is the so-called Harned coefficient. This is specific to the electrolytes i and j, the ions of which are interacting with each other, and depends on the ionic strength [7]. This dependence is more pronouned at low concentrations, and becomes very moderate at $I > 1$ mole/l.

The activity coefficient of neutral species (nonelectrolytes) are governed by Setchenov's equation for salting out (or in): $\log y_s = \log y_s^\circ + \Sigma_j k_{js} c_j$. Here y_s° is the activity coefficient of the nonelectrolyte in the absence of electrolytes j, k_{js} is the Setchenov (interaction) coefficient for the nonelectrolyte s and the electrolyte j. This relationship usually holds up to a few mole l^{-1}, but in concentrated solutions of electrolytes the amount of water bound to them by hydration (hydration number h) must also be taken into account. This effect diminishes the amount of water available for the solution of the nonelectrolyte, thus raising its activity coefficient. Salting out (positive k_{js}) is the rule, and only rarely is salting in observed. This may occur with electrolytes having ions of low charge and large size, that may interact attractively with the nonelectrolyte [8].

Electrolytes are usually supposed to be completely dissociated to their constituent ions in aqueous solutions. (Weak electrolytes, such as phosphoric acid, can be treated in terms of deprotonation equilibria, i.e., definite chemical reactions, and not only the physical interactions that are discussed in this section). However, in anticipation of the discussion of the behaviour of electrolytes in organic solutions, association of the ions of moderately strong electrolytes (e.g. 2:2 ones, such as $CuSO_4$) in aqueous solutions must also be discussed. This asssociation is governed by electrostatics, according to Bjerrum's theory. If α is the fraction dissociated to ions, then the association constant is given by $K_{ass} = (1-\alpha)/c\alpha^2 y_{i\pm}^2$, where $y_{i\pm}$ is the mean ionic activity coefficient of the dissociated part. (The activity coefficient of the neutral, associated, part is approximated as unity.) The value of K_{ass} is given by theory as $\log K_{ass} = 3 \log |z_+ z_-| + 6.120 - 3 \log \varepsilon + \log Q(b)$. Here ε is the relative permitivity (dielectric constant) of the solvent, $Q(b)$ is a (numerically) evaluatable integral function of the variable b, which is given by $\log b = \log |z_+ z_-| + 1.746 - \log \varepsilon - \log a$. The parameter a, given in nm, is the distance of nearest approach of the ions and characterizes the associating electrolyte. Thus, if ε of the solvent and a, z_+, and z_- of the electrolyte are known, b, $Q(b)$, K_{ass} and α can, in turn, be evaluated, the last mentioned iteratively, since $\log y_\pm = -A|z_+z_-|(\alpha I)^{\frac{1}{2}}(1 + Ba(\alpha I)^{\frac{1}{2}})^{-1} + b(\alpha I)$ is the mean ionic activity coefficient that applies to the dissociated part [9].

In addition to ionic association in the organic phase that increases as ε diminishes, and can be evaluated in terms of the Bjerrum theory as shown above, other interactions in the organic phase must also be taken into account. These are often due to dispersion forces between neutral solute and solvent molecules, and are then governed by 'regular solution' theory. According to this, the particles are distributed at random near each other, i.e., $\Delta S^E = 0$, there being no excess of entropy above the ideal entropy of mixing. The activity coefficient of a neutral species S is then given by $\log f_S = K_S(T)(1 - x_S)^2$, where $K_S(T)$ is a temperature- (but not composition-) dependent parameter [10].

Species in the organic phase are often aggregated, but aggregation alone, if it proceeds ideally, cannot lead to phase separation. However, if the aggregating species interact by dispersion forces so as to produce regular-solution behaviour with a sufficiently large positive value of K_S, then phase separation [11] (or 'third-phase formation' in distribution systems) may occur. Phase separation will occur if ΔG^E, the excess of molar Gibbs free energy of the system is sufficiently positive and large, and this is given by $\Sigma_j x_j RT \ln f_j$. On the contrary, if the interactions of a solute with the solvent are larger than with itself, a negative value of ΔG^E results, negative deviations from Raoult's law are observed, and the solution is stable against phase separation.

The aggregation of a solute in the organic phase, whether an electrolyte or a highly polar neutral species in a nonpolar organic solvent, proceeds according to the equilibrium requirement $\mu = \mu^\circ + RT \ln c + RT \ln y = \mu^\circ + RT \ln c_1 + RT \ln y_1$, where subscript $_1$ denotes the monomeric species and absence of a subscript the nominal component. This relationship leads to a value for the concentration of the monomer: $c_1 = cy/y_1$. The value of y is obtainable by means of vapour pressure osmometry, from the Gibbs–Duhem relationship. If, now, ideal aggregation is assumed, i.e., that the activity coefficient of the actual species (monomers and oligomers) of the solute are unity, $y_1 = 1$ and $c_1 = cy$ can be evaluated. The mean

aggregation number $\bar{n} = c / \int_0^c (1/y) dc_1$ can now be calculated as a function of the concentration c of the solute [12].

Once the activity coefficients of all the species that participate in the extraction equilibrium have been evaluated, it is possible to calculate the conditional extraction equilibrium constant K'_{ex} from the dependence of the distribution ratio D on the concentrations of all the reacting species. The value of K'_{ex} is 'conditional', since it depends on the nature and concentration of the non-reacting species that help determine the activity coefficients. When the conditions are kept constant, it is possible to evaluate the temperature dependence of K'_{ex} in terms of the enthalpy and entropy changes of the extraction reaction.

Some typical extraction reactions can be quoted that provide examples of some types of non-ideal distribution systems. Lithium chloride–water–n-hexanol is a three component system, in which the distribution can be described in terms of the reaction

$Li^+(aq) + Cl^-(aq) \rightleftharpoons Li^+(org) + Cl^-(org) \rightleftharpoons Li^+Cl^-(org)$. This equilibrium is shifted to the right to a readily measurable extent only at relatively high concentrations (several mole l^{-1}). The presence of other metal chlorides (such as magnesium or aluminium chlorides) shifts this equilibrium further to the right, both by the common-ion effect, but also by their dehydrating effect, which manifests itself by increased activity coefficients of the lithium chloride in the aqueous phase. Since hexanol and water have some mutual solubility, in particular water in hexanol, changes in the water content of the organic phase must also be taken into account [13].

Another typical distribution system is the ternary system uranyl nitrate–water–tri-n-butyl phosphate (TBP). This can be extended into a quaternary system by the addition of a diluent—kerosene or dedecane — to the organic phase, and to a quinquenary system by the further addition of nitric acid to the aqueous phase. The extraction reaction is $UO_2^{2+}(aq) + 2NO_3^-(aq) + 2TBP(org) \rightleftharpoons UO_2(NO_3)_2 \cdot TBP_2(org)$. The conditional extraction constant is given by $K'_{ex} = D \cdot [NO_3^-]^{-2}[TBP]^{-2}y_{\pm}^{-3} (\bar{y}_{solvate}/\bar{y}^2_{TBP})$. For the lack of better information the last factor, i.e. the ratio of activity coefficients in the organic phase, is sometimes considered to be a constant that can be incorporated into the extraction constant to give $\log K''_{ex} = \log D - 2 \log [NO_3^-] - 3 \log y_{\pm} - 2 \log [TBP]$. If this holds true (depending on the diluent, if present), then 'slope analysis' should give a constant value of 2 for $(\partial \log D/\partial \log [TBP])$, the partial derivative denoting a constant aqueous phase. If nitric acid is added, then $y_{\pm}$ for uranyl nitrate now denotes the quantity in the mixed aqueous solution. Another effect of the nitric acid, of course, is to provide the common nitrate anion. However, two further effects must be taken into account: one is the enhancement of the extraction by the association of the uranyl cation and the nitrate anion to form the species $UO_2NO_3^+(aq)$. The other is the reduction in the extent of extraction due to the competition of the nitric acid for the available TBP, forming the adduct $HNO_3 \cdot TBP$. These two reactions can be studied independently of the extraction of uranium, and equilibrium constants can be assigned to them, to be incorporated appropriately in the expression for K'_{ex}. Also, the activity coefficient $\bar{y}_{TBP}$ in the organic phase can be evaluated independently (say, by vapour phase osmometry or gas chromatography), so that only $\bar{y}_{solvate}$ must be evaluated from the distribution data for uranium [14].

A third example is the substitution reaction $UO_2^{2+}(aq) + 2 (HX)_2 (org) \rightleftharpoons UO_2X_2(HX)_2(org) + 2H^+(aq)$, where HX is di(2-ethylhexyl)phosphoric acid (HDEHP). Note that, at moderate acidities, the anion present (e.g., nitrate or perchlorate) does not play a role in the extraction. Also note that the extractant is a dimer in a diluent such as toluene. The main complication that occurs in this system is the aggregation of the extracted uranyl di(2-ethylhexyl)phosphate, solvated by two further HDEHP units, to larger aggregates. This makes the distribution ratio dependent on the concentration of the uranium. If TBP is added to the system it acts as a synergist by releasing HDEHP molecules from adduct formation, making them available for further extraction. An excess of TBP acts as an antagonist by itself binding the HDEHP. In this extraction system the aqueous phase is relatively simple (although it is perforce a mixed electrolyte: $UO_2A_2(aq) + HA(aq)$, A^- being the anion), but the organic phase is complicated, and the activity coefficients of the various species it contains are difficult to evaluate [15].

2.3 ENTHALPY AND ENTROPY CONTROL OF EXTRACTION EQUILIBRIA

The standard molar enthalpy and entropy changes of distribution equilibria yield information concerning the interactions that take place in the system beyond that obtainable from (isothermal) distribution curves alone. However, in order to make use of these quantities they must be determined in a reliable manner.

The slope of a (linear) plot of log K'_{ex} against the reciprocal of the absolute temperature should be a measure of the standard molar enthalpy of the extraction reaction $(\Delta H'_{ex} = -RT^2(d \ln K'_{ex}/dT) = -R(\ln 10) (d \log K'_{ex}/d(1/T)))$. If instead of the conditional extraction equilibrium constant K'_{ex} the extraction quotient Q_{ex} is used in this expression, is the slope still a measure of this enthalpy change? (The extraction quotient Q_{ex} is the ratio of the products of the concentrations of the extraction products divided by the product of the concentrations of the extraction reactants). The answer is no, since even if the activity coefficients or their ratio do not vary appreciably with the concentrations of the reactants, they are still temperature dependent [16]. It is, in general, not permissible to neglect the temperature variation of Y_{ex}, the activity coefficient quotient of the extraction reaction $(K'_{ex} = Q_{ex}Y_{ex})$. The temperature derivative of an activity coefficient is a heat of dilution or a heat of mixing. These quantities can often be measured independently, and taken into account.

When the extraction is due to a single well defined reaction, then calorimetry is the preferred method for the determination of the standard molar enthalpy change of the extraction. The heat change Δq is measured when Δn moles of distribuend transfer from the aqueous phase to the organic phase. (In stripping, Δn will be negative.) In order to evaluate the standard molar enthalpy of the extraction reaction ΔH°_{ex}, from this measured molar heat change, $\Delta q/\Delta n$, corresponding to a change from some initial state i to the final equilibrium state f, certain adjustments are necessary. In a general way this can be expressed as

$$\Delta q = \Delta n \cdot \Delta H^\circ_{ex} + \sum \int_i^0 \Delta H_{dil} \, d(\Delta n') + \sum \int_0^f \Delta H_{dil} \, d(\Delta n') \ ,$$

where the summation extends over all the species which have concentrations that change during the extraction. The quantity ΔH_{dil} symbolizes a heat of dilution or of mixing, and $\Delta n'$ symbolizes the change in the number of moles of the species involved [16].

Once the standard molar enthalpy change, ΔH°_{ex}, is obtained and the corresponding standard molar Gibbs free energy change is evaluated from the extraction equilibrium constant, $\Delta G^\circ_{ex} = -RT \ln K_{ex}$, the standard molar entropy change can be calculated from $\Delta S^\circ_{ex} = (\Delta H^\circ_{ex} - \Delta G^\circ_{ex})/T$. It is necessary to specify the standard state symbolized by the superscript $^\circ$. This may be defined as the state of infinite dilution of all the reactants in pure water and pure organic solvent (if their mutual solubility is negligible) or in mutually saturated water and organic solvent. This state may be designated by an asterisk. Alternatively, if the organic phase involves an extractant in a diluent then the standard state can be defined as involving also infinite dilution of this extractant, and is designated by superscript $^\infty$. Occasionally it is convenient to employ an inert electrolyte in the aqueous phase (e.g., $NaClO_4$) as a constant ionic medium, making the activity coefficients of the reactants and products independent of their concentrations, provided these are small compared with that of the inert electrolyte. In that case the standard state for the aqueous phase is the constant concentration of this inert electrolyte in water, it being understood that it does not distribute measurably into the organic phase. Quantities pertaining to this set of standard states should be marked accordingly. As a generalization of the various sets of standard states the superscript ° will serve [17].

Once the standard molar enthalpy and entropy of the extraction reaction have been evaluated, it may be asked which, if either, of these has the greater influence on the extraction equilibrium. Enthalpy control obtains if $|\Delta H^\circ_{ex}| > |T\Delta S^\circ_{ex}|$, and then the enthalpy change is always negative for successful extraction in the forward direction of the reaction. However, cases are known where entropy control obtains, and then $|\Delta S^\circ_{ex}| > |\Delta H^\circ_{ex}|/T$. In such cases control may mean a positive ΔS°_{ex}, which favours the extraction, but it may also mean a negative ΔS°_{ex}, which constitutes a barrier to the extraction in the standard state. Actual extraction then takes place only under conditions far from the standard state. For stripping or reverse extraction (from the organic to the aqueous phase) these relationships are, of course, reversed [2].

As an example of an enthalpy-controlled extraction system that of uranyl nitrate extraction by TBP can be cited. In the system $UO_2(NO_3)_2$–water–TBP–dodecane, $\Delta H^\infty_{ex} = -55$ kJ/mole, compared with $T\Delta S^\infty_{ex} = -9$ kJ/mole [16]. It is instructive to analyse the energetics leading to the observed standard molar heat of extraction.

The following hypothetical sequence of reactions is followed. The uranyl cation is removed from its aqueous environment into the gas phase and thereby dehydrated: $UO_2^{2+}(aq) \rightarrow UO_2^{2+}(g)$. The amount of enthalpy that has to be invested is $+1361$ kJ/mole. Two nitrate anions are similarly dehydrated: $2NO_3^-(g) \rightarrow 2NO_3^-(g)$, with an investment of 2×324 kJ/mole. The cation and the anions associate in the gas phase: $UO_2^{2+}(aq) + 2NO_3^-(g) \rightarrow UO_2(NO_3)_2(g)$, liberating electrostatic association energy to the extent of -1642 kJ/mole. This gaseous neutral species then associates with two molecules of gaseous TBP, forming bonds between the phosphoryl oxygen and the uranium atoms: $UO_2(NO_3)_2(g) + 2TBP(g) \rightarrow UO_2(NO_3)_2TBP(g)$, releasing a further amount of 2×202 kJ/mole. Finally, the diluent effects, due to the vaporization of two moles of TBP from, and condensation of the solvate into, dodecane are responsible for a release of a further -18 kJ/mole. It is seen that the net effect of this hypothetical sequence is the measured standard molar enthalpy change $\Delta H^\infty_{ex} = -55$ kJ/mole for the extraction reaction $UO_2^{2+}(aq) + 2NO_3^-(aq) + 2TBP(dod) \rightarrow UO_2(NO_3)_2(dod)$. This is only a small fraction of the total enthalpy changes involved. Of course, extraction does not take place via the gas phase, but the ions do get dehydrated, they also do associate electrostatically (the solvate in dodecane is not dissociated ionically), and TBP molecules do bond to the uranium atom. So whatever the actual mechanism, the net effect of the hypothetical sequence of steps is indeed observed.

We may now speculate on the effects of changes in the extraction system on its energetics, and through the enthalpy control on the distribution of the uranium. If TBP is exchanged for a better extractant, say trioctylphosphine oxide, TOPO, with a more basic phosphoryl group, the enthalpy released per bond is increased from

202 kJ/mole to some larger value. This can be estimated from the relative shift of the infrared vibration band of the P=O group in the two extractants in the presence of uranium. The addition of a non-extractable nitrate, such as $Al(NO_3)_3$, will cause association to $UO_2NO_3^+$. The dehydration of this univalent cation consumes much less energy than that of the divalent cation UO_2^{2+}, and only one nitrate anion must be dehydrated. The net effect of this drastic reduction in the energy that has to be invested for dehydration outweighs appreciably the loss in electrostatic association energy in the gas phase of $UO_2NO_3^+$ with the second nitrate anion. These examples indicate the considerations that can lead to useful results from the analysis of the standard molar enthalpy change of the extraction [17].

Another example of enthalpy-controlled extraction is that of the uranyl cation by HDEHP. The measured standard molar enthalpy change $\Delta H_{ex}^{\infty} = -30$ kJ/mole is appreciably larger, in the absolute sense, than $T\Delta S_{ex}^{\infty} = +7$ kJ/mole. The reaction to which these values pertain is $UO_2^{2+}(aq) + 2(HDEHP)_2(dod) \rightarrow UO_2(DEHP)_2(HDEHP)_2(dod) + 2H^+(aq)$, i.e. it is an exchange of the uranyl cation for two hydrogen ions of two HDEHP dimers [18]. The latter is cyclic, with two hydrogen bonds, and two such cycles, with one hydrogen ion (and the corresponding hydrogen bond) removed form chelate rings with the uranyl cation. The anions in the aqueous phase play no role at all in this extraction.

A hypothetical sequence of steps can, again, be used to describe the reaction. The uranyl cation is dehydrated, requiring an investment of $+1361$ kJ/mole. The dissociation of two hydrogen ions and their replacement by the uranyl cation in isolated HDEHP molecules in the gas phase is endothermic, requiring the further investment of $+797$ kJ/mole. (This includes small contributions from the diluent effects on removal of 2 $(HDEHP)_2$ from the dodecane and their replacment by the solvate.) The invested energy is returned with 'interest', when the two hydrogen ions released in the gas phase are returned to the aqueous phase and rehydrated there: -2×1094 kJ/mole are released. The net effect of this sequence of steps is, again, the measured standard molar enthalpy of the extraction reaction [17].

An example of entropy-controlled extraction is that in the system tetraphenylarsonium bromide–water–propylene carbonate (PC). The two solvents have a wide miscibility gap at room temperature, but the data shown below pertain to transfer of the solute from its standard state in pure water to its standard state in pure propylene carbonate. Since this latter solvent has a high relative permittivity (dielectric constant), the solute is expected to be completely dissociated in it, so that the transfer reaction is: $Ph_4As^+(aq) + Br^-(aq) \rightarrow Ph_4As^+(PC) + Br^-(PC)$. The standard molar entropy for the transfer is $\Delta S_{ex}^* = +18$ JK^{-1}mole^{-1}, whereas $\Delta H_{ex}^*/T = +8$ JK^{-1}mole^{-1} [19]. The positive entropy change thus overcomes the positive enthalpy barrier to this transfer. The solvation of the cation by itself does not provide an enthalpy barrier: on the contrary, the energy that has been expended in producing the cavity for the large cation in the highly structured solvent water is released on transfer to the less structured propylene carbonate. But this favourable enthalpy change is more than negated by the non-ability of the dipolar aprotic solvent carbonate to solvate the anion Br^-, which is well hydrated in the aqueous phase. The favourable entropy change which decides the direction of the transfer can be ascribed to the breakdown of the hydrophobically produced ice-like structure of the water around the large cation when this leaves the aqueous solution.

A final example of an entropy-controlled extraction system is one where the entropy change constitutes a barrier to the extraction. The system is lithium bromide–water–2-ethylhexanol, and the measured quantities are $\Delta S_{ex}^{\infty} = -127\,JK^{-1}mole^{-1}$ and $\Delta H_{ex}^{\infty}/T = -27\,JK^{-1}mole^{-1}$ only. Thus under the standard state condition, infinite dilution, extraction in the forward direction $Li^+(aq) + Br^-(aq) \rightarrow Li^+Br^-(org)$ is strongly hindered. (If saturated solutions are taken as the standard states, the barrier to extraction is lowered to $\Delta S_{ex}^{sat} = \Delta H_{ex}^{sat}/T = -94\,JK^{-1}mole^{-1}$ [20]. What is the source of this large entropic barrier to the extraction? It turns out that since the 2-ethylhexanol is saturated with water, the hydration of the lithium ion is not affected appreciably when it is extracted. However, in the organic phase the hydrated lithium ion is ion-paired with the bromide ion, so that there is a loss of translational entropy of $-13\,JK^{-1}mole^{-1}$. Another loss of entropy occurs when the water-structure-breaking bromide anion leaves the aqueous phase and the structure of bulk water is re-established: $-97\,JK^{-1}mole^{-1}$. This combined loss of entropy is only partly balanced by the entropy gained when water molecules are released from the hydration shell of the lithium cations and bromide anion, to be replaced by fewer alcohol moelcules and water molecules that together solvate the ion pair: $+101\,JK^{-1}mole^{-1}$. The net effect, therefore, is an entropy barrier to extraction. That this barrier is lowered in the saturated solutions is clear, since then the water has no structure left, so that the part of the entropy loss due to water-structure effects is no longer operative [17].

It must be remembered that the analysis given in the above four examples of the standard molar enthalpy or entropy of extraction in terms of the contributions of (hypothetical) steps is just a series of conjectures. What remain solid facts are the (calorimetrically) measured enthalpies together with the equilibrium constants for the extraction reaction. These permit the statement of the direction and extent of the extraction under the standard conditions, and its temperature dependence. Together with the necessary activity coefficients and the concentrations of the relevant reactants, the extent of extraction (the distribution ratio D) for any of the conditions covered by the treatment can be interpolated (and, with care, extrapolated) and predicted.

This is, thus, the purpose of the thermodynamic consideration of extraction reactions.

REFERENCES

[1] Y. Marcus and A. S. Kertes, *Ion Exchange and Solvent Extraction of Metal Complexes*, Wiley-Interscience, London, 1969, p. 437.
[2] Y. Marcus, *Introduction to Liquid State Chemistry*, Wiley Interscience, Chichester, 1977, p. 136.
[3] Ref. 1, p. 451.
[4] Ref. 2, p. 184.
[5] Ref. 1, p. 54.
[6] Ref. 2, p. 236.
[7] Ref. 1, p. 63, ref. 2, p. 247.
[8] Ref. 1, p. 74.
[9] Ref. 1, p. 189, ref. 2, p. 241.
[10] Ref. 2, p. 167, 190.
[11] Ref. 2, p. 149.
[12] Ref. 2, p. 184, ref. 1, p. 451.
[13] Y. Marcus and T. Nakashima, *Hydromet.*, 1982, **9**, 135.

[14] Y. Marcus, *J. Phys. Chem.*, 1961, **65**, 1647; ref. 1, p. 899.
[15] Ref. 1, p. 836; Y. Marcus and Z. Kolarik, *Inorg. Nucl. Chem. Lett.*, 1974, **10**, 275.
[16] Y. Marcus and Z. Kolarik, *J. Chem. Eng. Data*, 1973, **18**, 155.
[17] Y. Marcus, *CIM Spec. Publ.*, 1979, **21**, (Proc. ISEC '77), 154.
[18] Y. Marcus and Z. Kolarik, *J. Inorg. Nucl. Chem.*, 1976, **38**, 1069.
[19] Y. Marcus, *Pure Appl. Chem.*, 1985, **57**, 1103.
[20] Y. Marcus, *J. Chem. Eng. Data*, 1975, **20**, 141.

3

Kinetics of solvent extraction

Henry Frieser

For almost the past 25 years, and mostly under NSF sponsorship, we have systemati-
cally studied the kinetics of extraction of metal chelates. Our work has been fruitful,
leading to the conclusion that the rate-determining step(s) involve chelate formation
in the aqueous phase. This in turn permitted the development of a simple and widely
applicable method for studying the kinetics and mechanism of metal chelate
formation, uniquely capable of application to those chelates of which the low water
solubility renders other methods useless.

3.1 IMPROVED EXTRACTION APPARATUS

Although our earlier experimental extraction kinetics setup [1, 2] was capable of
yielding reliable data, it was somewhat inconvenient and, moreover, limited to
systems in which the reaction half-life was at least 3 min. The apparatus developed
during the current grant period, based on the high-speed (up to 20000 rpm) stirrer
and the Morton flask was found to improve greatly the precision of the data,
particularly when half lives were short. An extraction with $t_{1/2}=100$ sec can be
described with a precision (standard deviation) of 1.4% [3]. This apparatus proved
vital to our studies of copper extraction and to reliable evaluation of activation
parameters in the nickel–dithizone systems.

More recently, we have developed a computer-controlled automated extraction
apparatus [4] in which the high-speed stirring apparatus was modified to utilize a
small Teflon phase-separator filter through which the organic phase was selectively
drawn at a rate of 5.0 ml/min. The organic solution was pumped through a 10-μl flow-
through spectrophotometric cell and then returned to the reaction mixture in the
flask by a peristaltic pump. In kinetic runs, the absorbance at the appropriate
wavelength was recorded as a function of time on an X–Y recorder and also
interfaced by an A/D converter to a Nova 2/10 Minicomputer programmed to control
the experiment and to obtain, analyse, and store the data. The data (usually a
thousand points were taken over at least two half-lives), were examined to validate

the assumption of pseudo-first order kinetics, and the rate constant was calculated along with an estimate of its reproducibility (standard deviation of slope and intercept). A test of the system with nickel and zinc dithizone extractions gave rate constants that not only closely matched earlier experimental values but which also had significantly smaller standard deviations. The time required to complete the experiment and analyse the data was of course dramatically decreased from several hours to as little as 10 minutes.

3.2   METAL–LIX SYSTEMS

We have studied [5] the extraction of Cu^{2+} by the commercial, high molecular weight aromatic hydroxyoxime, LIX65, and demonstrated that, contrary to earlier opinion that the kinetics were 'interfacially controlled'' (Appendix C), the slow step is the aqueous phase reaction:

$$CuL^+ + HL \xrightarrow{k_1} CuL_2 + H^+$$

We also found that the commercial aliphatic hydroxyoxime, LIX63, catalyses the LIX65 extraction according to one of the following equations:

$$CuL^+HB \rightarrow CuLB + H^+$$

or

$$CuB^+ + HL \rightarrow CuLB + H^+$$

where HB represents LIX63.

We made a similar investigation of the Cu(II)–LIX63 system [63]. Under the conditions we used, only the simple 1:2 Cu–LIX63 complex was detected. It may be that the use of chloroform as solvent was unfavourable for the unusual 1:1 Cu complex with dianionic LIX63. The rate of extraction was found to be first order in ligand, and the overall kinetic behaviour serves to clarify the mechanism of the LIX63 catalysis of the LIX65 extraction as:

$$Cu^{2+} + HB \rightarrow CuB^+ + H^+ \quad fast$$
$$CuB^+ + HL \rightarrow CuBL + H^+ \quad slow$$

As a further test of the validity of the 'aqueous phase' mechanism, the kinetics of the Cu–LIX65 extraction was studied for six additional organic solvent systems [7]. The change in the apparent reaction rate should depend only on the aqueous-phase LIX65 concentration, which in turn is inversely proportional to K_{DR}, the distribution constant. It was gratifying to observe that, throughout the seven systems, the apparent rate contants varied in excellent quantitative agreement with our mechanism (inverse-square dependence as demanded by the second-order rate dependence on the ligand). A parallel study was made for the Ni–LIX65 system and here, where there is a first order dependence on ligand, a good linear plot of the apparent rate constant *vs.* the reciprocal of K_{DR} was obtained. Thus, although our findings do not support the 'aqueous' mechanism for all conditions, there can be little doubt of its validity under our experimental conditions. Indeed, the question of whether, or rather where, the mechanism changes as the hydrophobicity is increased further, is very significant for both fundamental and practical aspects of extractant design principles (see Section 3.5).

3.3 NICKEL–DITHIZONE ACTIVATION PARAMETERS

Earlier work on the role of dithizone substituents on the rate of formation of 1:1 nickel and zinc complexes clearly demonstrated that both electron-releasing and withdrawing substituents increased the rate constant [8]. The increase seemed to correlate roughly with the size of the substituent, however, and this suggested that the volume swept out by the ligand (and hence the entropy of the formation) might play a leading role. Accordingly, we studied the temperature dependence of the rate of formation of nickel chelates of dithizone and its methylated and halogenated derivatives [9]. In all the systems studied, the activation entropies were unusually high and negative, clustering around -218.8 ± 3.0 e.u. and the activation enthalpies unusually low, from 4.4 to 0.0 kcal/mole. These interesting results confirm the importance of the entropy factor, but also point to the possibility of an additional, energetically favourable, reaction step that, from analogy with literature data, may have a $\Delta H°$ of around 10 kcal/mole and a $\Delta S°$ of as much as -30 e.u.

These effects are mutually compensating, explaining the 'normal' $\Delta G^\dagger$ and rate constants observed. Such good compensation might make one look askance at the validity of the observations, but the theme (almost equal and opposite enthalpy and entropy effects) is a fairly general one. In any event, we scrupulously examined our procedures and are convinced that our findings are real. The considerable differences we observe between the kinetic behaviour of dithizone and that observed by more conventional techniques with other ligands strongly suggests something unique about this ligand. Recently, we found spectrophotometric evidence of a fairly strong chloroform–dithizone complex [10]. The complex, in cyclohexane solution, has the formula $HDz.2CHCl_3$, with a formation constant of $10^{4.0}$. Inasmuch as this constant does not appear to have any appreciable temperature coefficient, it would not seem to help explain the unusual activation enthalpies described above.

3.4 BACK EXTRACTION KINETICS

The kinetics of back extraction (or stripping) of metal chelates is interesting from a number of fundamental and practical aspects. When there is possibility of a change in mechanism (from homogeneous to interfacial) of extraction occurring as the ligand becomes increasingly hydrophobic (higher K_D), then examination of back extractions of metal chelates, inherently more hydrophobic than the corresponding ligands (usually), and a study of their mechanisms, should greatly assist in solving this question. Typically, metal chelates are stripped by acidic solutions, reversing the extraction reactions. It is therefore expected that the back extraction rate would be catalysed by H_3O^+. If this is the case, as was indicated in preliminary experiments, it becomes interesting to explore the comparative effects of other acids, most notably Lewis acids such as strongly complexing metal ions. Does the back extraction proceed through acid attack of the chelate, or does it first dissociate? Further, can Lewis bases (i.e. other ligands) catalyse back extraction, particularly if the chelate is co-ordinatively unsaturated?

An interesting series of chelates to use for such a study is the Ni(II) dithizonates which have long been known to exhibit slow rates of formation in extraction. Also, the pH required for the back extraction is several units below that needed for extraction; the effect is a kind of 'hysteresis'. The experimental method closely followed that used in this laboratory for extraction kinetics. The rates of back extraction were followed by monitoring the concentration of the nickel chelate in the chloroform phase spectrophotometrically. In every case, the back extractions were found to be first order in chelate [11]. Experimental rate expressions were elaborated by observing the dependencies of the pseudo-first-order constants on the relevant concentration variables. In some cases, only qualitative comparisons of rates could be obtained. Our results show that reagents affecting back-extraction rates can have two distinct levels of effectiveness. Among the acids, Ag^+ and Hg^{2+} enhance back extraction rates much more than does H_3O^+. Cyanide is a much better stripping catalyst than EDTA. One possible way to explain this is that the effective agents from an intermediate complex with the chelate which can rapidly dissociate while the others involve an inherently slow dissociation mechanism. The intermediate complex involves bonding with either the central metal ion (provided that its co-ordination number under these conditions is capable of increasing) or with one of the bonding atoms of the ligand (provided that it has unused bonding capability). In the systems studied here, cyanide, a softer ligand than EDTA, probably adds to the nickel ion in much the same fashion as do neutral nitrogen ligands [12]. In fact, the rates of both extraction and back extraction of the mixed-ligand chelate nickel–dithizone–phenanthroline ($NiDz_2$.phen) are much greater than those for nickel–dithizone [13]. Both Ag^+ and Hg^{2+} can bond with S (even thioether S) to a much greater extent than can Cu^{2+} (which doesn't catalyse stripping of $NiDz_2$) and probably interact by doing so.

3.5 ROLE OF THE INTERFACE IN SOLVENT EXTRACTION

The recent introduction of the Teflon phase separator filter [4] permitted us to see for the first time dramatic changes in the extent of extraction under 'equilibrium conditions' that occurred when the two phase system was stirred at high speed. Under such conditions, a significant interfacial area is obtained which makes it possible to observe the effects of adsorption into the interfacial region. The phenomenon can be illustrated by the behaviour of diphenylthiocarbazone (dithizone) and its alkylated analogues (di-p-alkylphenylthiocarbazones). The extraction behaviour of these lipophilic weak acids is characterized by the value of $pH_{1/2}$, the pH at which 50% extraction occurs (a measure of K_D/K_a, the ratio of distribution constant to acid dissociation constant) obtained from the variation in distribution ratio D, with pH in the alkaline range. Although essentially no difference in $pH_{1/2}$ is observed with dithizone between phase separation under high-speed stirring and under the more conventional conditions of no stirring during separation, increasingly large changes were observed with the size of the alkyl substituent. Thus, $pH_{1/2}$ values of 0.16, 0.67, 3.0 and 5.0 were observed with methyl, ethyl, n-butyl, and n-hexyl groups, respectively [14].

We developed expressions describing the variation of D with the interfacial volume (a function of the stirring rate and of the specific configuration of the stirring flask) and pH, based on a scheme involving distribution equilibria of (a) the neutral ligand between the two bulk phases and a third, interfacial phase and (b) the ligand anion between the bulk aqueous phase and the interfacial phase. Measurements of the $CHCl_3$–water interfacial tension revealed that, although at low pH values (e.g. 3.5) relatively no change occurred when butyldithizone was dissolved in the chloroform, at high pH values (e.g. 11.0) the interfacial tension dropped significantly with increasing butyldithizone concentration in $CHCl_3$. This would seem to indicate that the butyldithizone anion, but not the neutral ligand, is surface active. These interfacial tension data gave, with the help of the Gibbs equation, information concerning the minimum concentration required for saturation of the interface, a parameter we would expect to decrease with increasing size of the alkyl substituent. We also found that K'_L, the distribution constant of the dithizone anion between the interfacial and bulk aqueous phases, another measure of the 'excess concentration' of the interfacially adsorbed species, did in fact increase with increasing size of the alkyl substituent. Finally, an examination of the kinetics of the extraction of both nickel and zinc by the alkyldithizones showed that the mechanism must be interfacial [15]. Interestingly enough, after the unusually high alkyldithizonate concentration in the interface had been acounted for, the kinetics of formation of the 1:1 MDz^+ chelate in the interface was not dramatically different from that observed in the bulk aqueous phase.

By means of our new-found ability to examine extraction systems under conditions of vigorous agitation rather than having to limit our observations to 'quenched' systems in which agitation ceased before phase separation took place, we not only have a significantly improved method of examining the kinetics and mechanisms of metal chelates, but have opened a new window on examination of unsupported liquid–liquid interfaces in general.

REFERENCES
[1] C. B. Honaker and H. Freiser, *J. Phys. Chem.*, 1962, **66**, 127.
[2] B. E. McClellan and H. Freiser, *Anal. Chem.*, 1964, **36**, 2262.
[3] S. P. Carter and H. Freiser, *Anal. Chem.*, 1979, **51**, 1100.
[4] H. Watarai, L. Cunningham, and H. Freiser, *Anal. Chem.*, 1982, **54**, 2390–2392.
[5] S. P. Carter and H. Freiser, *Anal. Chem.*, 1980, **52**, 511.
[6] V. Bagreev and H. Freiser, *Sep. Sci. Technol.*, 1982, **17**, (5), 751.
[7] K. Akiba and H. Freiser, *Anal. Chim. Acta*, 1982, **136**, 329.
[8] J. S. Oh and H. Freiser, *Anal. Chem.*, 1967, **39**, 295.
[9] K. Ohashi and H. Freiser, *Anal. Chem.*, 1980, **52**, 767.
[10] S. P. Bag and H. Freiser, *Anal. Chim. Acta*, 1982, **136**, 439.
[11] K. Ohashi and H. Freiser, *Anal. Chem.*, 1980, **52**, 2214.
[12] K. S. Math and H. Freiser, *Anal. Chem.*, 1962, **41**, 1682.
[13] B. S. Freiser and H. Freiser, *Talanta*, 1970, **17**, 540.
[14] H. Watarai and H. Freiser, *J. Am. Chem. Soc.*, 1983, **105**, 191.
[15] H. Watarai and H. Freiser, *J. Am. Chem. Soc.*, 1983, **105**, 189.

4

From Nernst to synergism

Erik Högfeldt
Department of Inorganic Chemistry, The Royal Institute of Technology, S-100 44 Stockholm, Sweden

4.1 THE EXTRACTION METHOD IN EQUILIBRIUM ANALYSIS

Extraction methods find wide application in analytical chemistry [1]; and technological applications found in hydrometallurgy and liquid ion exchange were discussed at recent conferences [2, 3]. This paper covers the use of extraction methods in equilibrium analysis, with particular reference to speciation, i.e. the complexes formed under particular experimental conditions.

4.1.1 The distribution law

Consider the compound A, distributed between an aqueous and an organic phase. The mutual solubility of the two phases should be negligible. If the state of aggregation of A does not change, the reaction can be written:

$$A(aq) \rightleftharpoons A(org) \tag{4.1}$$

The law of mass action applied to reaction (1) gives

$$\{A\}_o / \{A\} = \lambda \qquad (p_{tot}, T \text{ const}) \tag{4.2}$$

$\{A\}$ = activity of A and λ = distribution coefficient of A. Equation (4.2) can be written in terms of concentrations giving

$$[A]_o / [A] = \lambda y / y_0 \approx \lambda_0 \tag{4.3}$$

Here y and y_0 are the activity coefficients of A in the two phases. If the ratio y/y_0 is practically constant, Eq. (4.3) reduces to Nernst's distribution law.

Often the same concentration scale is used in both phases, but this is not necessary. Experimentally, the stoichiometric distribution coefficient is determined, i.e. the ratio of the total concentrations of A in the two phases. This quantity, D, is given by

$$D = [A]_o^{tot}/[A]^{tot} \tag{4.4}$$

Determination of D as a function of $[A]$ or some related variable gives data that can be treated by the methods of equilibrium analysis.

4.1.2 The extraction of water and acids by aromatic hydrocarbons
In practical extraction, complex hydrocarbon mixtures are often used. Several criteria are important, such as good phase separation, good solubility of the components involved, solvent not too inflammable, etc. These diluents are often regarded as inert, but water and acids are extracted by most diluents. Sometimes even metals are extracted [4,5]. Below, some examples will be given of the application of equilibrium analysis to water and acid extraction.

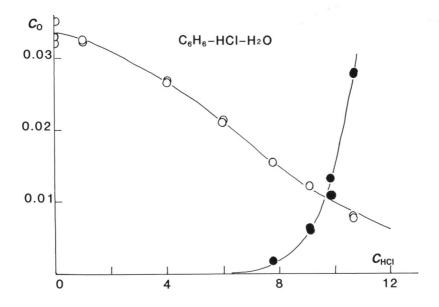

Fig. 4.1 — The concentration in the organic phase, C_o, plotted against the equilibrium concentration of HCl in the aqueous phase, C_{HCl}, for the system: C_6H_6–HCl–H_2O. Concentration unit M except for water in the aqueous phase, where the mole fraction scale is used. $T = 298$ K. Data from [6].

○ $[H_2O]_o$ ● $[HCl]_o$

The curves were computed with the constants given in Table 4.1.

4.1.2.1   The system C_6H_6–HCl–H_2O

In Fig. 4.1, $[A]_o$ is plotted against the equilibrium concentration of HCl, C_{HCl}, in the aqueous phase for benzene shaken to equilibrium with HCl–H_2O mixtures [6]. It is evident that water and acid are extracted independently of each other.

If the two species are assumed to behave ideally in the organic phase, the following expressions are obtained for the distribution coefficients

$$[H_2O]_o/\{H_2O\} = \lambda_1 \qquad [HCl]_o/\{H^+\}\{Cl^-\} = \lambda_2 \qquad (4.5a,b)$$

The solubility of benzene in water is so low that the activities of water and acid in pure HCl–H_2O mixtures can be used. Table 4.1 gives values for $\log\lambda_1$ and $\log\lambda_2$ obtained from Eqs. (4.5a,b) with data for some aromatic hydrocarbons [6]. The curves in Fig. 4.1 were computed with the constants given for benzene in Table 4.1.

Table 4.1 — Distribution coefficients λ_1 and λ_2 for aromatic hydrocarbons–HCl–H_2O. $T = 298$ K. Data from [6].

Hydrocarbon	$\log \lambda_1$	$\log \lambda_2$
Benzene	$- 1.470 \pm 0.014$	$- 6.63 \pm 0.07$
Toluene	$- 1.581 \pm 0.009$	$- 6.63 \pm 0.04$
m-Xylene	$- 1.689 \pm 0.004$	$- 6.63 \pm 0.03$
o-Xylene	$- 1.673 \pm 0.007$	$- 6.66 \pm 0.05$
p-Xylene	$- 1.611 \pm 0.012$	$- 6.61 \pm 0.01$

NMR measurements indicate that at most $\sim 3\%$ of the water in benzene might be aggregated, probably to dimers [7]. It has been shown elsewhere [8] that such a low extent of aggregation can be neglected.

These results show that most of the water is monomeric in aromatic hydrocarbons, contrary to opinion at that time [9].

4.1.2.2   The system nitrobenzene–water

The extent of aggregation of water varies from one diluent to another. In nitrobenzene, there is a considerable shift of the water protons, indicating aggregation of water in this diluent. By shaking nitrobenzene with LiCl–H_2O and CaCl$_2$–H_2O mixtures, the water activity could be varied. In Fig. 4.2, D is plotted against $\{H_2O\}$ for those measurements [7]. The water in the organic phase was (as for the other hydrocarbons) determined by the Karl Fischer method. D is given by

$$\begin{aligned} D &= [H_2O]_o^{tot}/\{H_2O\} \\ &= [H_2O]_o/\{H_2O\} + 2[(H_2O)_2]_o/\{H_2O\} \\ &= \lambda + 2K_2\lambda^2\{H_2O\} \end{aligned} \qquad (4.6)$$

λ is the distribution coefficient of the monomer and K_2 the dimerization constant. Disregarding the lowest water activity studied, a straight line can be fitted to the experimental data. The constants in Table 4.2 were obtained by linear regression.

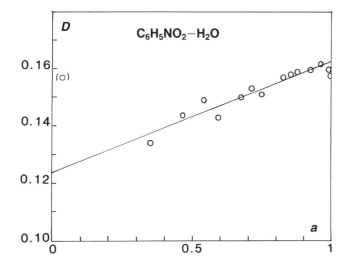

Fig. 4.2 — The distribution coefficient D, defined by Eq. (4.6) plotted against $\{H_2O\} = a$ for the system $C_6H_5NO_2$–H_2O. The water activity was defined by LiCl or CaCl$_2$ solutions. $T = 298$ K. Data from [7]. Concentration scales as for Fig. 4.1. The straight line was computed from the constants in Table 4.2, by linear regression.

Table 4.2 — The system $C_6H_5NO_2$–H_2O. $T = 298$ K. Data from [7]

Choice	Method of computation	λ_1	K_2	U	$\sigma(\log D) \times 10^3$
1	Average D	0.153 ± 0.008	0	3.54×10^{-4}	± 5.03
2	Linear regression	0.124	1.23	2.02×10^{-5}	± 2.02
3	LETAGROP	0.124 ± 0.003	1.20 ± 0.20	2.15×10^{-5}	± 2.12

Neglecting aggregation, the quantity $\lambda_0 = D$ was also computed and given in Table 4.2. Also the constants obtained by using the computer program LETAGROP with the same data are given in Table 4.2. In order to get a measure of the goodness of fit, the error-squares sum, U, was computed. This quantity is defined by

$$U = \sum_{1}^{n} (\log D_{exp} - \log D_{calc})^2 \quad (n = \text{number of experimental points}) \quad (4.7)$$

Table 4.2 also gives $\sigma(\log D)$ given by

$$\sigma \log D) = \pm \sqrt{U/(n-1)} \quad (4.8)$$

That slightly different constants are obtained is due to the fact that in LETA-GROP the minimization was done for $[H_2O]_o^{tot}$. As seen from Table 4.2, a considerable improvement in the fit to the experimental data is obtained by taking aggregation into account.

In Fig. 4.3, $[H_2O]_o^{tot}$ is plotted against the equilibrium molarity of sulphuric acid,

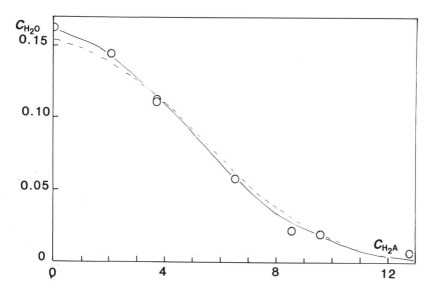

Fig. 4.3 — C_{H_2O} plotted against C_{H_2A} for the system $C_6H_5NO_2$–H_2SO_4–H_2O. $T = 298$ K. Data from [10].
O experimental points. Concentration scales the same as for Fig. 4.1.
————— Computed from choice 2 in Table 4.2.
– – – – – Computed from choice 1 in Table 4.2.

$C_{H_2SO_4}$, for the system $C_6H_5NO_2$–H_2SO_4–H_2O [10]. The co-extraction of sulphuric acid is negligible below $10\,M$ H_2SO_4. The continuous curve was computed from choice 2 in Table 4.2, using the water activities given by Giauque *et al.* [11] for the system H_2SO_4–H_2O. For comparison, the dashed curve was computed from choice 1 in Table 4.2. For $C_{H_2SO_4} \geqslant 2$ the fit is not too bad. The importance of the additional information from the NMR data is obvious.

4.1.2.3   The system $C_6H_5NO_2$–$HClO_4$–H_2O

The extraction of water and perchloric acid by nitrobenzene has been studied by Högfeldt *et al.* [12]. Both phases were analysed for acid and the organic phase for water, and NMR, conductivity and viscosity measurements were made on the organic phase.

In Fig. 4.4 the concentration of water in the organic phase, C_{H_2O}, is plotted against the equilibrium molarity of perchloric acid in the aqueous phase, C_{HClO_4}, and in Fig. 4.5 the acid concentration in the organic phase, $C_{HA(o)}$ is plotted against

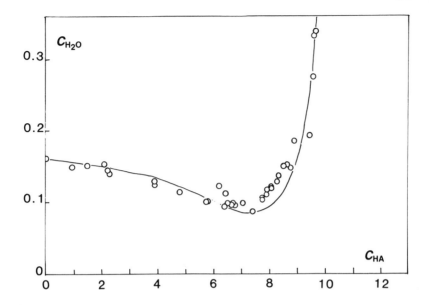

Fig. 4.4 — C_{H_2O} plotted against C_{HA} for the system $C_6H_5NO_2$–$HClO_4$–H_2O. $T = 298$ K.
Concentration scales the same as for Fig. 4.1. Data from [12].
○ Experimental points. The curve was computed from the constants in Table 4.3.

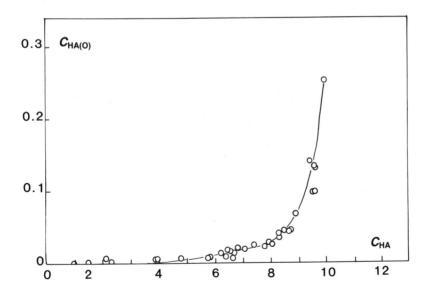

Fig. 4.5 — $C_{HA(o)}$ plotted against C_{HA} for the system $C_6H_5NO_2$–$HClO_4$–H_2O. $T = 298$ K. Data
from [12]. Concentration scales as for Fig. 4.1.
○ Experimental points. The curve was computed from the constants in Table 4.3.

C_{HClO_4}. It is evident that mixed acid–water complexes are formed in the organic phase, because the water concentration starts to increase when acid starts to be extracted. Observe the slow rise of $C_{HA(o)}$ in the range 4–8 M HClO$_4$. This implies appreciable dissociation in the organic phase as verified in Fig. 4.6 where $\kappa \times 10^5$ is

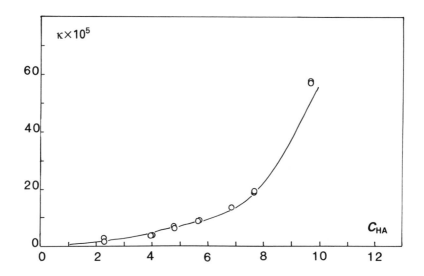

Fig. 4.6 — $\kappa \times 10^5$ plotted against C_{HA} for the system C$_6$H$_5$NO$_2$–HClO$_4$–H$_2$O. $T = 298$ K. Data from [12]. Concentration scales as for Fig. 4.1.
○ Experimental points. The curve was computed from the constants in Table 4.3.

plotted against C_{HClO_4}. In Fig. 4.7 the number of water molecules per acid molecule in the organic phase $(\bar{n})$ is plotted against C_{HClO_4}. In computing $\bar{n}$ the water extracted by nitrobenzene itself was corrected for by use of the constants of choice 3 in Table 4.2. Although the spread is considerable, $\bar{n}$ seems to approach 2 at high acidities.

In order to get information about the water content of the conducting species the following approach was used. Consider the reaction

$$H^+ + ClO_4^- + nH_2O \rightleftharpoons H(H_2O)_{n_+}^+ + ClO_4(H_2O)_{n_-}^-$$ (4.9)

$$n_+ + n_- = n$$

Application of the law of mass action to this reaction gives

$$2\log C_n = \log K_n + \log\{H^+\}\{ClO_4^-\} + n\log\{H_2O\}$$

$$C_n = [H(H_2O)_{n_+}^+] = [ClO_4(H_2O)_{n_-}^-]$$

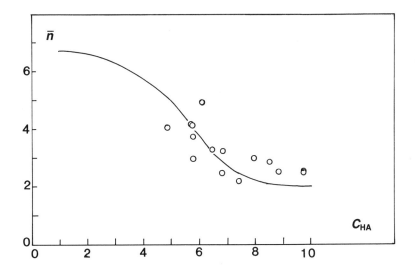

Fig. 4.7 — The number of water molecules per acid molecule, $\bar{n}$, plotted against C_{HA} for the system $C_6H_5NO_2$–$HClO_4$–H_2O. $T = 298$ K. Data from [12]. The curve was computed from the constants in Table 4.3.

and

$$\log C_n = \tfrac{1}{2}\log K_n + \tfrac{1}{2}\log\{H^+\}\{ClO_4^-\} + \frac{n}{2}\log\{H_2O\} \qquad (4.10)$$

Between C_n and the conductivity, κ, the following expression is assumed to hold:

$$\kappa = 10^{-3}C_n l_n \qquad (4.11)$$

l_n is the equivalent conductivity of the ion-pair under consideration. It is assumed that l_n is constant throughout the range where the ion-pair with n water molecules makes a noticeable contribution to the conductivity. From Eqs. (4.10 and 4.11)

$$\log\kappa - \tfrac{1}{2}\log\{H^+\}\{ClO_4^-\} = -3.00 + \tfrac{1}{2}\log k_n + \log l_n + \frac{n}{2}\log\{H_2O\}$$

or

$$\log Y = \text{const} + \frac{n}{2}\log\{H_2O\}$$

In Fig. 4.8 $\log Y$ is plotted against $\log\{H_2O\}$. Observe that at low water activities

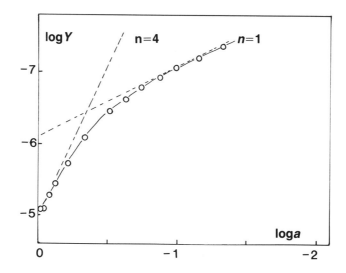

Fig. 4.8 — Log $\kappa - \frac{1}{2}\log\{H^+\}\{ClO_4^-\} = \log Y$ plotted against $\log\{H_2O\} = \log a$ for the system $C_6H_5NO_2$–$HClO_4$–H_2O. $T = 298$ K. Data from [12].
Two asymptotes are given, one with a slope of unity, the other with a slope of 4. The curve was computed from the constants in Table 4.3.

the slope tends towards unity, indicating that at high acidities the conducting species contain two water molecules. This is supported by Fig. 4.7. The simplifying assumption that l_n is practically constant thus receives some support from the analytical data. Table 4.3 gives the final model, obtained by applying various

Table 4.3 — The system $C_6H_5NO_2$–$HClO_4$–H_2O. $T = 298$ K. Data from [12]

Species HA:H_2O	log K	l	δ(ppm)
[1:8	$-6.46(-5.36)$*	16.56 ± 0.19	-0.15 ± 0.51]
1:2	-7.74 ± 0.03	6.80 ± 0.08	-4.24 ± 0.19
3:6	-17.94 ± 0.07	0	-3.71 ± 1.27

The chemical shifts (δ) are relative to H_2O(l).
The equivalent conductance (l) is defined by Eq. (4.11).
*Here log $(K + 3\sigma(K))$ is given. In the other cases log $K \pm 3\sigma(\log K)$.

versions of LETAGROP to the analytical, conductance and NMR data. An earlier model arrived at graphically was used as input for the computer calculation. In addition to a species with 2 water molecules, another with more water is needed, as seen in Fig. 4.8. An asymptote with $n = 4$ (i.e. 8 water molecules) can be fitted to the data. However, this conducting hydrate is most uncertain, so the constants in Table 4.3 referring to it are given in brackets. One nonconducting species containing two water molecules is required. The steep rise of the curve in Fig. 4.5 at high acidities suggests this species to be aggregated. The best fit was obtained with $(HClO_4(H_2O)_2)_3$. The curves in Figs 4.4–4.8 were computed from the constants in Table 4.3. Although only tentative, the model gives a rather satisfactory description of a complicated system.

4.1.3 Metal extraction
Two examples taken from the literature will be used to illustrate the use of metal extraction for investigating complex formation in aqueous solution.

4.1.3.1 The hydrolysis of vanadium(V) at low concentrations
One advantage of the extraction method is the possibility of working at very low concentrations by using tracer methods. This can be illustrated by the study of the hydrolysis of vanadium(V) by Dyrssen and Sekine [13].

In Fig. 4.9, log$D$ is plotted against pH. The temperature is 298 K and the ionic

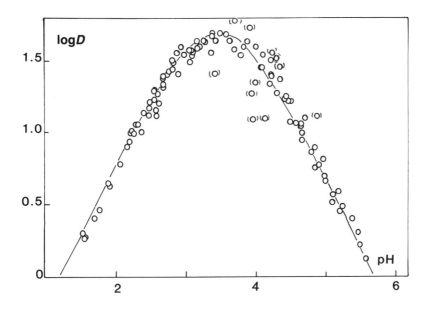

Fig. 4.9—Log D, defined by Eq. (4.13), plotted against pH for the system vanadium(V)–0.5 M (Na,H)ClO$_4$. The curve was computed from constants obtained graphically (Table 4.4). Data from [13].

strength in the aqueous phase kept at 0.5 M by (Na,H)ClO$_4$. The extractant was methylisobutylcarbinol (hexanol). The tracer was carrier-free ^{48}V. The extractable species HVO$_3$ can take up one proton or lose one proton. These equilibria can be written

$$VO_2^+ + 2H_2O \rightleftharpoons H_3O^+ + HVO_3 \qquad (K_1) \qquad\qquad (4.12a)$$

$$HVO_3 + H_2O \rightleftharpoons H_3O^+ + VO_3^- \qquad (K_2) \tag{4.12b}$$

For the distribution of HVO_3 the reaction can be written

$$HVO_3 \rightleftharpoons HVO_3(o) \qquad (\lambda) \tag{4.12c}$$

The expression obtained for the stoichiometric distribution coefficient is:

$$D = \frac{[HVO_3]}{[VO_2^+] + [HVO_3] + [VO_3^-]} \tag{4.13}$$

This can be normalized by introducing

$$v = \sqrt{K_1 K_2}/h \qquad p = \sqrt{K_1/K_2} \qquad h = [H_3O^+] \tag{4.14a,b}$$

giving

$$Q = D/\lambda p = 1/(v + p + v^{-1}) \tag{4.15}$$

where p is a shape parameter. The plot $\log Q(\log v)_p$ has two asymptotes

$$\underset{v \to \infty}{\log Q = -\log v} \qquad \underset{v \to 0}{\log Q = \log v}$$

The two asymptotes meet at

$$\log Q = 0 = \log D - \log \lambda p \qquad log\ v = 0 = \tfrac{1}{2}\log K_1 K_2 - \log h \tag{4.16a,b}$$

A set of curves $\log Q(\log v)_p$ is computed for different p-values. These are then plotted on transparent paper, on the same scale as the experimental curve $\log D(pH)$. The set of normalized curves is placed over the experimental curve. In the position of best fit there is one curve $\log Q(\log v)_p$ that has the same shape as the experimental curve. This gives p. From the intersection point of the asymptotes the three constants defined by reactions (4.12a–c) are obtained by using equations (4.16a,b). In Table 4.4 the constants arrived at in this way are compared with those

Table 4.4 — A comparison of estimates of the hydrolysis constants of vanadium(V) at low concentrations. $T = 298$ K. Ionic strength $= 0.5$ M (Na,H)ClO$_4$. Data from [13].

Method	$\log K_1$	$\log K_2$	$\log \lambda$	U	$\sigma(\log D)$
Graphical [13]	-3.3	-3.7	2.1	0.350	± 0.058
Graphical This work	-3.33	-3.68	2.07	0.338	± 0.057
LETAGROP [13]	-3.203 ± 0.030	-3.783 ± 0.029	1.989 ± 0.027	0.283	± 0.052

given in the original paper. The two graphical estimates are in fairly good agreement. The constants obtained by LETAGROP differ slightly, and as expected they give an improved fit to the experimental data.

The curve in Fig. 4.9 was computed from the constants obtained graphically in this paper. The fit is satisfactory in view of the large spread in the experimental data. (This is partly compensated for by the large number of experimental points.) The points within parentheses in Fig. 4.9 were not used in the computation of U and $\sigma(\log D)$ in Table 4.4.

In view of the strong tendency of vanadium(V) to form polymeric species, it would have been difficult to study the mononuclear hydrolysis by other methods.

4.1.3.2 Complex formation between Th(IV) and acetylacetone

Rydberg [14] studied the complex formation between Th(IV) and acetylacetone (HA) in water by distributing HA and ThA$_4$ between benzene and an aqueous phase, with the ionic strength kept at 0.01 M by (Na,H)ClO$_4$ at 298 K. In Fig. 4.10 $\log D$ is plotted against $\log [A^-] = \log a$. The expression for Q becomes

$$Q = D/\lambda_4 = \beta_4 a^4 / (1 + \beta_1 a + \beta_2 a^2 + \beta_3 a^3 + \beta_4 a^4) \qquad (4.17)$$

Besides the distribution coefficient of ThA$_4$ (λ_4) there are four stability constants to be determined for the species ThA^{3+} to ThA$_4$. Rydberg estimated four independent stability constants, whereas Dyrssen and Sillén [15] used a two-parameter approach to fit the same data.

In the following, a number of two-parameter approaches will be applied to Rydberg's data.

(1) *The Dyrssen–Sillén two-parameter equation.* The following two parameters were introduced

$$\log \beta_N = NA \qquad \log(K_n/K_{n+1}) = 2B \qquad (4.18a,b)$$

K_n is the nth stepwise formation constant and β_N is the stability constant for the highest complex formed under the experimental conditions used, usually corresponding to the characteristic co-ordination number.

The normalized variable v is introduced:

$$v = \beta_4^{1/4} \times a \qquad (4.19a)$$

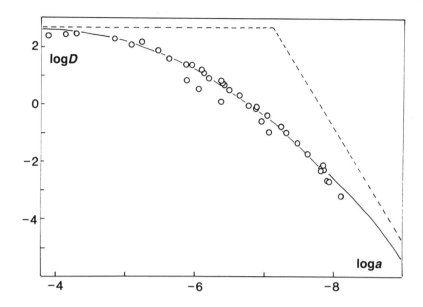

Fig. 4.10 — Log D, defined by Eq. (4.17), plotted against log $[A^-] = $ log a for the system Th(IV)–HA–C_6H_6–0.01 M (Na,H)ClO$_4$. $T = 298$ K. Data from [14]. The curve was computed from set no. 3 in Table 4.5.

which together with Eqs. (4.18a,b) gives

$$Q = D/\lambda_4 = v^4/(1 + v \times 10^{3B} + v^2 \times 10^{4B} + v^3 \times 10^{3B} + v^4) \qquad (4.19b)$$

The constant B is the shape parameter, like p in Eq. (4.15). The two asymptotes log $Q = 0$ and log $Q = 4$ log v intersect at log $v = 0 = \frac{1}{4}$ log $\beta_4 +$ log a $= A +$ log $a(v = 1)$ i.e. $A = -$ log $a(v = 1)$. The intersection point thus gives A, and B is obtained from the curve of the set of normalized curves log $Q(\log v)_B$ that gives the best fit with the experimental curve. In the original paper a rather elaborate way for estimating B from a single point was suggested.

Table 4.5 lists the stability constants estimated in this paper, along with those originally suggested by Rydberg as well as by Dyrssen and Sillén. Before the results in Table 4.5 are discussed another two-parameter equation will be used with the data.

(2) *The Bjerrum spreading factor.* Jannik Bjerrum [16] introduced the spreading factor defined by

$$K_n/K_{n+1} = f_n x^2 \qquad (4.20)$$

where f_n is the statistical ratio between two stepwise stability constants and x is an unknown constant, the spreading factor. For $x = 1$ the statistical ratios apply; f_n is given by

Table 4.5 — Comparison of methods for determining the stability constants for the reaction between Th(IV) and acetylacetone in 0.01 M (Na,H)ClO$_4$. $T = 298$ K. Original data from [14].

Set	Investigator or method	log β$_1$	log β$_2$	log β$_3$	log β$_4$	log λ$_4$	U	σ(log D)
1	Rydberg [14]	7.85	15.58	21.85	26.86	2.50	1.90	±0.22
2	Dyrssen–Sillén [15]	8.01	15.12	21.33	26.64	2.43	4.32	±0.33
3	Dyrssen–Sillén	9.52	17.45	23.76	28.48	2.65	1.72	±0.21
	This work							
4	Spreading factor	8.43	15.83	22.28	27.68	2.15	2.82	±0.27
	This work							
5	LETAGROP DISTR (1)	—	15.74	22.04	26.86	2.54	1.57	±0.20
	This work							
6	LETAGROP DISTR (2)	8.55	16.37	22.73	27.53	2.55	1.63	±0.20
	This work							
7	Average	8.47 ± 0.65	16.02 ± 0.81	22.33 ± 0.84	27.34 ± 0.69	2.47 ± 0.17	1.83	±0.22

$$f_n = \frac{(n+1)(N-n+1)}{n(N-n)} \tag{4.21}$$

where N is the same quantity as in Eq. (4.18a).

Using the same normalized variable as before [Eq. (4.19a)], the following expression is obtained for Q:

$$Q = D/\lambda_4 = v^4/(1 + 4x^3v + 6x^4v^2 + 4x^3v^3 + v^4) \tag{4.22}$$

From Eq. (4.22) it is seen that x is the shape parameter. From Eq. (4.19a) β_4 is obtained, and with the aid of Eq. (4.21) all the stability constants can now be obtained. The results of fitting normalized graphs log $Q(\log v)_x$ to the data in Fig. 4.10 are given in Table 4.5.

(3) *The Mihailov equation.* Recently Mihailov [17] suggested the following equation for the overall stability constant β_n

$$\beta_n = BA^n/n! \tag{4.23}$$

With the normalized variable v defined by Eq. (4.19a) the expression for Q becomes

$$Q = D\lambda_4 = v^4/(1 + 2.21B^{3/4}v + 2.45B^{1/2}v^2 + 1.81B^{1/4}v^3 + v^4) \tag{4.24}$$

The constant B is now the shape parameter. From the intersection of the two asymptotes

$$\log v = \tfrac{1}{4}\log \beta_4 + \log a(v = 1)$$

giving

$\frac{1}{4}$log $B/24$ + log A = − log $a(v = 1)$

However, it was impossible to fit the data in Fig. 4.10 with Eq. (4.24). The discussion will thus be limited to a comparison of the other two approaches with that obtained by LETAGROP DISTR, all given in Table 4.5.

Computer calculations. The program LETAGROP DISTR for fitting distribution data [18] was used in order to find out how far from the best possible fit are the constants estimated graphically.

The data fed into the computer were pH, $[HA]^{tot}$, $[Th(IV)]^{tot}$, $V_o/V(= 1)$ and D, the distribution coefficient defined by Eq. (4.17). In the calculations two different choices for ligand variable were used: (1) $[HA]_{org}$, (2) $[HA]_{aq}$. Starting values for the constants were obtained from the stability constants suggested by Rydberg [14].

Discussion. In order to compare Sets 1–6 in Table 4.5 on the same basis, the constants obtained by the various methods were for the computer results transformed into stability constants, β_n, directly obtained by the other methods. Using the estimates of log $[A^-]$ = log a, given by Rydberg [14], log D-values were computed and U and $\sigma(\log D)$ computed from Eqs (4.7) and (4.8).

From Table 4.5 it is evident that, as expected, the best fit is obtained by using LETAGROP. Unfortunately, the stability constants obtained from (1), $[HA]_o$, differ considerably from those obtained from (2), $[HA]_{aq}$. This is due to the large spread in the experimental data as illustrated by the unusually large error-squares sums and the large uncertainty in log D as given by $\sigma(\log D)$.

Both sets 3 and 4 give a better fit than set 2, obtained by Dyrssen and Sillén. Unfortunately set 3 gives a set of constants widely different from the others. It must be regarded as a local, "false", minimum not representing the "true" minimum (Set 5, which has the lowest value for U.).

Set 7 gives the averages of the constants obtained in Sets 1–6. It seems that in treating poor data, simple graphical methods with few parameters give satisfactory results. If possible a few different two-parameter approaches should be tried. In this way the number of curve-fitting parameters is kept at a minimum.

4.2   THE PHYSICAL CHEMISTRY OF AMINE EXTRACTION

Salts of long-chain amines dissolved in a suitable diluent have, since the second world war, been used as liquid ion-exchangers in analytical chemistry, in hydrometallurgy, in the nuclear industry etc. Neverthless, amines are still attracting interest [2].

This section discusses the physical chemistry of such extractants with emphasis on equilibrium analysis.

4.2.1   The protonation reaction and the aggregation of the salt

4.2.1.1   The reaction
Consider the following reaction

$$A(\text{org}) + H^+ + X^- \rightleftharpoons AHX(\text{org}) \qquad (K_1) \qquad\qquad (4.25)$$

where an organic base dissolved in a suitable medium reacts with an acid in an aqueous phase to form a salt in the organic phase. The solubility of A and AHX in the aqueous phase can be regarded as negligible.

This simple protonation reaction can be studied by the two-phase emf titration

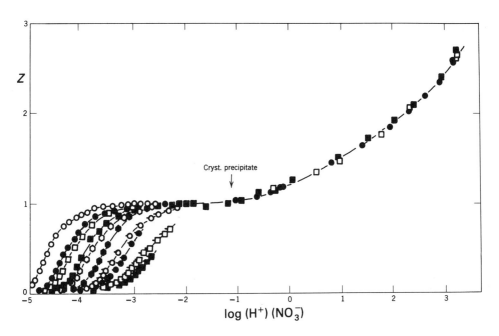

Fig. 4.11 — Z_{HNO_3} plotted against $\log\{H^+\}\{NO_3^-\}$ for $Z > 1$, and against $\log[H^+][NO_3^-]$ for $Z < 1$ for the system TLA–n-octane–HNO$_3$ for 10 amine concentrations. The ionic strength $I = 1\,M$ (Na,H)NO$_3$, $T = 298$ K. Data from [20]. Reprinted from A. S. Kertes and Y. Marcus, editors, *Solvent Extraction Research*, 1969, p. 159, by permission of the copyright holders, John Wiley & Sons, Inc.

method [19]. Figure 4.11 gives a plot of Z_{HNO_3} against $\log\{H^+\}\{NO_3^-\}$ for the extraction of nitric acid by trilaurylamine (TLA) dissolved in n-octane [20]. The number of acid molecules extracted by each amine molecule, Z_{HA}, is given by

$$Z_{HA} = ([HA]_{org}^{tot} - [HA]_{dil})/A \qquad\qquad (4.26)$$

Some acid might be extracted by the diluent at high acidities, and this must be subtracted from the total amount. A is the total concentration of amine in the organic phase.

It is convenient to divide the range of Z-values into two parts:

I: $0 \leqslant Z \leqslant 1$

II: $Z > 1$

Range I corresponds to reaction (4.25), the range where pure base is transformed into salt. This is the range that can be studied by the two-phase titration method. The data for this range in Fig. 4.11 were obtained by titrating an aqueous phase of ionic strength $I = 1\,M$ (Na,H)NO$_3$. In this range $\{H^+\}\{NO_3^-\}$ is replaced by $[H^+][NO_3^-] = [H^+] = h$ since $[NO_3^-] = 1$, and $[HNO_3]_{dil} = 0$. The data for $Z > 1$ were obtained by batch experiments in which both phases were analysed for acid and the organic phase for water by the Karl Fischer method.

Returning to the range $Z \leqslant 1$ and reaction (4.25) the following expression is obtained for Z if this reaction is the only process that needs to be considered:

$$Z_{HX} = K_1 ah/(a + K_1 ah) = K_1 h/(1 + K_1 h) \tag{4.27}$$

where a is the free amine concentration in the organic phase. According to Eq. (4.27), all the curves of $Z(\log h)_A$ should coincide. This is not the case in Fig. 4.11, where a set of curves is obtained, one for each amine concentration. This can be understood if the salt aggregates to form dimers, trimers etc.

In the range $Z > 1$ all the curves coincide, indicating that the processes occurring in this range seem to be independent of size. It is this range that is used in practical extraction processes, when acid anions are exchanged for various anionic complexes.

In the following, the aggregation equilibria in some simple amine systems will be treated by the methods of equilibrium analysis, especially the methods for studying the formation of polynuclear complexes.

4.2.1.2 $\bar{p},\bar{q}$-analysis
Consider the general reaction

$$pA + qB \rightleftharpoons A_p B_q \quad (K_{p,q}) \tag{4.28}$$

Sillén [21] has shown how the averages $\bar{p}$ and $\bar{q}$ are defined by

$$\bar{p} = \sum p[A_p B_q] \Big/ \sum [A_p B_q] = \frac{A - a}{S}$$

$$\bar{q} = \sum q[A_p B_q] \Big/ \sum [A_p B_q] = \frac{B - b}{S}$$

A and B are the total concentrations of the two components, a and b are the concentrations of the uncomplexed components. S is the complexity sum.

In Fig. 4.12, q is plotted against p in the range $p \leqslant 10, q \leqslant 8$, for some simple

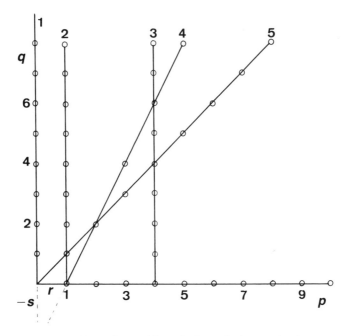

Fig. 4.12 — q plotted against p for p ≤ 10, q ≤ 8. 1, p = 0. Aggregation in one-component systems. 2, p = 1. Formation of mononuclear complexes. 3, Homonuclear complexes with p = 4. 4, Core + links complexes. 5, polynuclear complexes with p = q.

special cases. With the aid of $\bar{p}$ and $\bar{q}$ it might be possible to limit the range of p- and q-values that need to be taken into account.

The quantity R is needed. It is defined by

$$AR = a + S$$

R can be obtained from (cf. [21]).

$$R = \text{const} - \int \left(\frac{\partial Z}{\partial \log A} \right)_b d\log b = \text{const} + \int \left(\frac{\partial \log b}{\partial \log A} \right)_Z dZ \qquad (4.29a,b)$$

When applying this approach to data like those in Fig. 4.11, we can arbitrarily set b = h and A = the total amine concentration in the organic phase.

From the spacing of the curves Z(logb)A, the derivatives in Eqs. (4.29a,b) can be obtained. The horizontal difference between two curves in Fig. 4.11, divided by

$\Delta \log A = \log A_2 - \log A_1$ gives an estimate of $\left(\dfrac{\partial \log b}{\partial \log A} \right)_Z$. Similarly an estimate of

$\left(\dfrac{\partial Z}{\partial \log A} \right)_b$ is obtained from the vertical distance between two curves in Fig. 4.11. By doing this for the range of Z-values covered esperimentally the integrals in Eqs. (4.29a,b) can computed either by graphical or numerical methods.

The integration constant in Eqs. (4.29a,b) can be obtained as follows:

1: The data can be extrapolated down to concentrations where free A predominates, such that $a \approx A$, $S \approx 0$. Then

$$AR = A + 0 \quad \text{and} \quad R = R_0 = 1$$

2: The extrapolation is uncertain; it is better to start the integration from a point where only one complex is present, denoted by $A_P B_Q$.

The mass-balance condition for A gives together with the definitions of R and Z:

$$A = a + P[A_P B_Q] \quad AR = a + [A_P B_Q] \quad AZ_0 = Q[A_P B_Q]$$

Elimination of $[A_P B_Q]$ gives

$$R = R_0 = 1 - Z_0(P - 1)/Q$$

where Z_0 is the lowest Z-value studied and R_0 the corresponding R-value.

With knowledge of R the fraction of free A present can be computed. This quantity is defined by

$$\alpha_0 = a/A$$

and can be computed from (cf. [21]).

$$\ln \alpha_0 = \ln a - \ln A = \text{const} + R - \int_{A = \text{const}} Z\mathrm{d}\ln b \tag{4.30}$$

i.e. by integrating the curves $Z(\log h)_A$ in Fig. 4.11. If $R_0 = 1$ (i.e. the free amine predominates) the constant in Eq. (4.30) becomes -1.

If $R_0 = 1 - Z_0 (P - 1)/Q$, the expression for α_0 is

$$\alpha_0 = \alpha_{00} = 1 - PZ_0/Q$$

Taking the first term in the series expansion of $\ln \alpha_{00}$ we get

$$\ln \alpha_0 = -(1 + Z_0/Q) + R - \int_{A = \text{const}} Z\mathrm{d}\ln b$$

Knowledge of α_0 permits evaluation of $\bar{p}$ and $\bar{q}$ from

$$\bar{p}=\frac{A-a}{S}=\frac{A-a}{AR-a}=\frac{1-\alpha_0}{R-\alpha_0} \qquad \bar{q}=\frac{B-b}{S}=\frac{AZ}{AR-a}=\frac{Z}{R-\alpha_0} \qquad (4.31a,b)$$

The computations of $\bar{p}$ and $\bar{q}$ are conveniently performed by the computer program MESAK [22].

4.2.1.3 The system TIOA–CCl₄–HCl.

Vieux [23] studied the extraction of hydrochloric acid by tri-iso-octylamine (TIOA) dissolved in carbon tetrachloride at 298 K. The ionic strength in the aqueous phase was presumably kept at 1.00 M. In Fig. 4.13 Z_{HCl} is plotted against log h for that

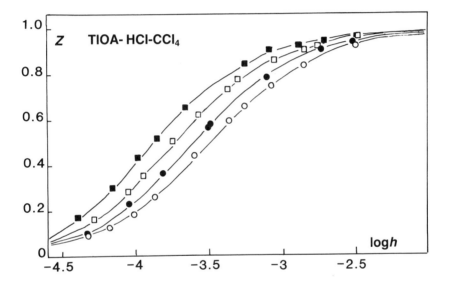

Fig. 4.13 — Z_{HCl} plotted against log h for the system TIOA–CCl₄–HCl. $T = 298$ K.
○ $A = 24.5$ mM ● $A = 54.2$ mM □ $A = 113.0$ mM ■ $A = 242$ mM
Data from [23].

system. Four amine concentrations ranging from 24.2 mM to 242 mM were used. The similarity to the curves in Fig. 4.11 suggests formation of aggregates.

In Fig. 4.14 $\bar{p}$ is plotted against $\bar{q}$ for the data in Fig. 4.13. The data cluster along the line $\bar{p}=\bar{q}$, i.e. line 5 in Fig. 4.12. This means that the predominating species have the 1:1 composition, i.e. they are $(TIOAHCl)_n$, neglecting possible co-extraction of water. Since $\bar{n}=\bar{p}=\bar{q}<2$ we start to consider a monomer–dimer equilibrium.

(a) *Use is made of* $\bar{p}=\bar{q}$. The fact that $\bar{p}=\bar{q}$ allows us to introduce the variable:

$$u = ah$$

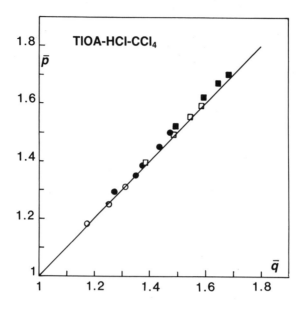

Fig. 4.14 — $\bar{p}$ plotted against $\bar{q}$ for the data in Fig. 4.13. The symbols refer to the same concentrations as in Fig. 4.13.

giving

$$AZ = K_1 u + 2K_2 u^2$$

$$A = a + K_1 u + 2K_2 u^2$$

From these equations

$$a = A(1 - Z)$$

and

$$u = Ah(1 - Z)$$

The expression for AZ can be rearranged to

$$AZ/u = K_1 + 2K_2 u$$

which can be compared with the normalized graph

$$Y = \frac{AZ}{uK_1} = 1 + y \qquad y = \frac{2K_2}{K_1}u$$

Figure 4.15 shows a plot of log AZ/u against log u for the data in Fig. 4.13. The

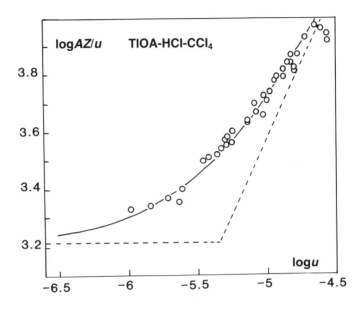

Fig. 4.15 — log AZ/u plotted against log u for the system TIOA–CCl$_4$–HCl, compared with the normalized curve in the position of best fit. The two asymptotes are dashed.

normalized curve log Y (log y) together with the two asymptotes in the position of best fit are also marked out. The asymptotes are

$$\log Y = \log AZ/u - \log K_1 = 0 \qquad \log Y = \log y = \log 2K_2/K_1 + \log u$$

The asymptotes intersect at log $Y = \log y = 0$. From the position of the horizontal

Table 4.6 — *The system TIOA–CCl$_4$–HCl. $T = 298$ K. Data from [23]*

Method	log K_1	log K_2	U	$\sigma(Z)$
Normalization log(AZ/u), log u	3.215	8.27	1.433×10^{-3}	± 0.0061
Projection map	3.23	8.26	1.399×10^{-3}	± 0.0061
LETAGROP ZETA	3.23 ± 0.04	8.26 ± 0.03	1.382×10^{-3}	± 0.0061

asymptote, $\log K_1$ ($= \log AZ/u_{Y=1}$) is obtained. From the point of intersection of the asymptotes $\log 2K_2/K_1 = -\log u_{y=1}$ is obtained. The values for $\log K_1$ and $\log K_2$ obtained in this way are given in Table 4.6 together with the residual-squares sum U and $\sigma(X)$ defined by

$$U = \sum_1^n (X_{exp} - X_{calc})^2 \quad (n = \text{number of experimental points}) \quad (4.32a)$$

(cf. Eq. (4.8).

$$\sigma(X) = \pm \sqrt{U/(n-1)} \quad (4.32b)$$

In Table 4.6 $X = Z$.

(b) *Projection map. Normalization of* $(AZ)_Z$. The following variables are introduced

$$u = K_1 h \qquad v = \left(\frac{K_2}{K_1}\right) a$$

giving

$$AZ = (K_1^2/K_2)\,(uv + 2u^2v^2)$$

$$A = (K_1^2/K_2)\,(v + uv + 2u^2v^2)$$

The normalized variables Y_1 and Y_2 are introduced

$$Y_1 = AZK_2/K_1^2 = uv + 2u^2v^2$$

$$Y_2 = AK_2/K_1^2 = v + uv + 2u^2v^2$$

These expressions give the following expression for Z

$$Z = (u + 2u^2v)/(1 + u + 2u^2v)$$

This equation can be used to compute v for a certain value of u and known Z. With u and v known, Y_1 and Y_2 can be computed and compared with experimental data, either A or AZ.

In Fig. 4.16, data for $\log AZ$ $(\log h)_Z$ are compared with the normalized curves in the position of best fit for $Z = 0.2, 0.4, 0.6$ and 0.8. From the position of the origin of the normalized graph K_1 is obtained from

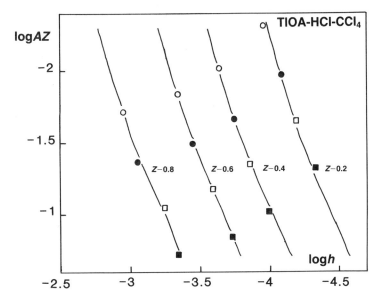

Fig. 4.16 — Projection map log $AZ(\log h)_Z$ compared with the normalized curves in the position of best fit for the system TIOA–CCl$_4$–HCl. The symbols are the same as in Fig. 4.13.

$$\log K_1 = -\log h(u = 1)$$

and K_2 from

$$\log(K_2/K_1^2) = -\log AZ(Y_1 = 1)$$

In Table 4.6 the constants obtained are given together with U and $\sigma(Z)$. The results obtained graphically were used as input data in the refinement of the constants by the program LETAGROP ZETA. The agreement between graphically estimated values and those obtained by computer is excellent.

In Fig. 4.17, the fractions of amine present as monomer, α_1, and as dimer, α_2, are plotted against log A. At the concentrations used in practical extraction processes, practically all the TIOAHCl is present as dimer.

In this treatment the possible co-extraction of water has been neglected. The water activity is kept constant by the ionic medium and can be included in the constants.

4.2.1.4 The concentration scale
The disparity in size between the large amine molecules and their salts, and the diluent, might suggest use of the volume fraction scale instead of mole fractions. In our studies and those of Vieux, the molarity scale is used. This scale is closely related to the volume fraction scale as shown below.

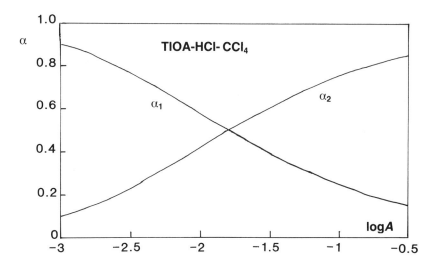

Fig. 4.17 — The fraction of amine present as monomer, α_1, and dimer, α_2, plotted against log A at $Z = 1$.

n_1 moles of A are mixed with n_2 moles of B and the molar volumes are v_1 and v_2. The volume fraction of A, ϕ_A is defined by

$$\phi_A = n_1 v_1 / (n_1 v_1 + n_2 v_2)$$

The total volume, V, is given by

$$V = n_1 v_1 + n_2 v_2$$

The molarity, C_A, is given by

$$C_A = n_1/V$$

and

$$\phi_A = v_1 C_A$$

The molarity and volume fraction scales are proportional to each other. If the volume of mixing is nonideal, the molar volumes have to be replaced by the partial molar volumes. For most organic systems the deviations from ideal mixing are small. It can be concluded that the molarity scale can be expected to be able to account for any disparity in size.

4.2.1.5 The activity coefficients in the organic phase

The excellent agreement between the model and the experimental data implies that consideration of the dimerization accounts for practically all deviations from ideality. Observe that the shapes of the curves $Z(\log h)_A$ and $\log A\,Z(\log h)_Z$ are predicted by the model. There are no adjustable parameters. The good agreement up to 240 mM solutions of TIOA in CCl$_4$ suggests that in other systems too, only aggregation equilibria need to be taken into account up to about 250 mM.

Marcus [24] has suggested that solute–solvent interactions are more important than solute–solute interactions, i.e. aggregation. However, since the diluent activity is practically constant, solute–solvent interactions can be included in the constants (K_n).

4.2.1.6 The system TLA–t–BB–HNO₃

The extraction of nitric acid by tri-n-dodecylamine=trilaurylamine (TLA) dissolved in tertiary butylbenzene (t-BB) was studied by Högfeldt and Fredlund [25]. The ionic strength in the aqueous phase was kept at 5.00 M by (Na,H)NO₃. In Fig. 4.18, $Z_{\mathrm{HNO_3}}$

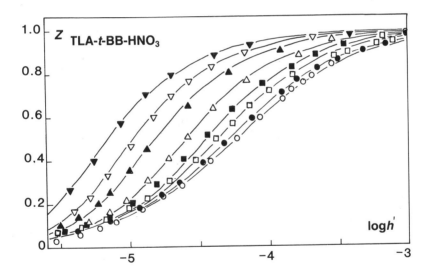

Fig. 4.18 — $Z_{\mathrm{HNO_3}}$ plotted against $\log h[\mathrm{NO_3}] = \log h'$ for the system TLA–t-BB–HNO₃. The ionic strength $I = 5$ (Na,H)NO₃. $T = 298$ K.

○ $A = 0.163$ mM	● $A = 1.67$ mM	□ $A = 4.03$ mM	■ $A = 6.96$ mM
△ $A = 15.73$ mM	▲ $A = 39.8$ mM	▽ $A = 75.45$ mM	▼ $A = 151.6$ mM

is plotted against $\log h$ for eight amine concentrations ranging from 0.163 mM to 151.6 mM. In Fig. 4.19 $\bar p$ is plotted against $\bar q$ for this system. Again, the data cluster along the 1:1-line indicates formation of (TLAHNO₃)$_n$, neglecting the co-extraction of water. Since $\bar n < 3$ it seems reasonable to start by considering formation of monomers and trimers only. First, use was made of the variable $u = ah$, and then the projection map.

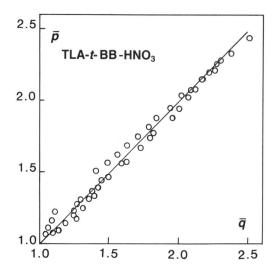

Fig. 4.19 — $\bar{p}$ plotted against $\bar{q}$ for the system TLA–t-BB–HNO$_3$.

(a) *Use of* $\bar{p} = \bar{q}$. The expression

$$AZ/u = K_1 + 3K_3u^2$$

was compared with

$$Y_1 = AZ/K_1u = 1 + y \qquad y = (3K_3/K_1)u^2$$

As before K_1 is obtained from the position of the horizontal asymptote and K_3 from the point of intersection of the two asymptotes. The constants obtained in this way are given in Table 4.7. Also U and $\sigma(Z)$ computed from Eqs. (4.32a,b) are given.
(b) *Projection map*. Here all three species with $n = 1, 2$ and 3 were considered giving

$$Y_1 = AZ\sqrt{K_3/K_1^3} = uv + 2\alpha u^2v^2 + 3u^3v^3$$

$$Z = (u + 2\alpha u^2v + 3u^3v^2)/(1 + u + 2\alpha u^2v + 3u^3v^3)$$

with

$$u = K_1h \qquad v = \sqrt{K_3 K_1^3} \times a \qquad \alpha = K_2/\sqrt{K_1K_3}$$

Curves were computed for from $\alpha = 0$ to $\alpha = 1$. In Fig. 4.20 log AZ is plotted against log h for $Z = 0.2, 0.4, 0.6$ and 0.8. The best fit was obtained with $\alpha = 0.5$. The curves in Fig. 4.20 are those for $\alpha = 0.5$ in the position of best fit.

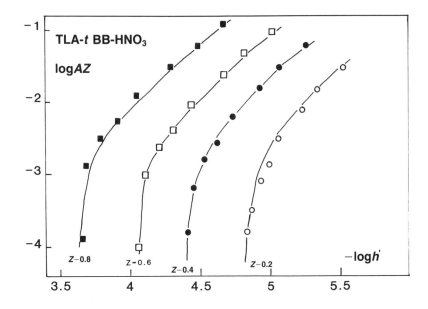

Fig. 4.20 — Projection map log $AZ(\log h')_Z$ compared with the normalized curves in the position of best fit. The symbols refer to the four Z-values used.

From the position of the origin of the normalized graph, K_1 is obtained from

$$\log y = 0 \qquad \log K_1 = -\log h \ (u = 1)$$

and K_3 from

$$\log Y = 0 \qquad \tfrac{1}{2}\log K_3 - \tfrac{3}{2}\log K_1 = -\log AZ \ (Y_1 = 1)$$

Knowing K_1 and K_3, K_2 is computed from the expression for α. The constants obtained in this way are given in Table 4.7 together with those computed from

Table 4.7 — The system: TLA–t-BB–HNO$_3$ $T = 298$ K Data from [25]

Method	$\log K_1$	$\log K_2$	$\log K_3$	U	$\sigma(Z)$
Normalization log (AZ/u), log u	4.28	—	17.20	2.54×10^{-2}	± 0.015
Projection map	4.22	10.36	17.10	2.16×10^{-2}	± 0.013
LETAGROP ZETA	4.235 ± 0.023	10.43 ± 0.21	17.00 ± 0.17	1.72×10^{-2}	± 0.012

LETAGROP ZETA. The constants found by using the projection map agree fairly well with those obtained by LETAGROP and the dimer is rather well established. Figure 4.21 shows the fraction of amine present in each species, α_i, plotted against log A. Although the dimer is never a predominant species, it is not negligible, ranging between 10 and 20% in the range of amine concentrations studied. At high amine concentrations the trimer predominates.

4.2.1.7 Influence of various factors on the aggregation process

Table 4.8 lists some systems studied by the two-phase titration method. The equilibrium constant K_n refers to a generalization of reaction (4.25)

$$nA(org) + nH^+ + nX^- \rightleftharpoons (AHX)_n(org) \qquad (K_n) \tag{4.33}$$

In some aliphatic diluents the following reaction sometimes seems to occur

$$2A(org) + H^+ + X^- \rightleftharpoons A_2HX(org) \qquad (K_{2,1}) \tag{4.34}$$

Table 4.9 gives the aggregation constants, k_n; these refer to the following process in the organic phase:

$$nAHX(org) \rightleftharpoons (AHX)_n(org) \qquad (k_n) \tag{4.35}$$

and is computed from

$$\log k_n = \log K_n - n\log K_1$$

In aromatic diluents, data at high amine concentrations indicate that large aggregates are formed: 30-mer, 50-mer etc. [30]. In aliphatic diluents they form over a more extensive range of amine concentrations, but their composition is uncertain. For that reason they are put in brackets in Table 4.8. If only data for concentrations below 150 mM are taken into account, the large species can be neglected for aromatic diluents. One reason for the difficulty in identifying the species formed is that at high acidities and amine concentrations the ligand variable $u = ah$ tends towards a constant value, [30], [40]. This 'constant' has been interpreted as a formal solubility product. Small variations in u makes resolution of the data into different species difficult. The precipitate formed in some aliphatic diluents might be regarded as the ultimate limit of the aggregation process.

That aggregation is the most important process to take into account can be inferred from Table 4.8. Besides the investigation by Vieux, where only two species need to be taken into account over a large range of amine concentrations there are a few others where only one species needs to be considered:

1 — Only monomer in the system TLA–o-Xylene–CCl$_3$COOH for 15.4 m$M \leqslant A \leqslant$ 161 mM.
2 — Only monomer in the system TLA–CHCl$_3$–HCl for 7.9 m$M \leqslant A \leqslant 147.3$ mM
3 — Only dimers in the system DINA–CHCl$_3$–HCl for 6.1 m$M \leqslant A \leqslant 295$ mM

Table 4.8 — Equilibrium constants for reactions (4.33) and (4.34) in various diluents. Data for the range $Z \leqslant 1$. $T = 298K$

Amine	Diluent	Acid	Ionic Medium	$\log K_n \pm 3\sigma \ (\log K_n)$	Ref.
TLA	o-Xylene	CCl$_3$COOH	1 (Na,H)X	$\log K_1 = 6.93 \pm 0.03$	26
THA	Benzene	HNO$_3$	1 (Na,H)X	$\log K_1 = 5.05 \pm 0.06$	27
				$\log K_2 = 11.73 \pm 0.08$	
TLA	Toluene	HNO$_3$	2.75 (Li,H)X	$\log K_1 = 5.02 \pm 0.04$	28
				$\log K_2 = 11.68 \pm 0.19$	
				$\log K_3 = 18.19(+ 0.21)*$	
TLA	m-Xylene	HNO$_3$	1 (Na,H)X	$\log K_1 = 4.26 \pm 0.02$	30 + 29
				$\log K_2 = 10.22 \pm 0.14$	
				$\log K_3 = 16.35 \pm 0.12$	
TOA	o-Xylene	HNO$_3$	1(Li,H)X	$\log K_1 = 4.62 \pm 0.02$	31
				$\log K_2 = 10.75 \pm 0.15$	
				$\log K_3 = 17.22 \pm 0.11$	
TLA	o-Xylene	HNO$_3$	1(Na,H)X	$\log K_1 = 4.50 \pm 0.03$	30 + 32
				$\log K_2 = 10.46 \pm 0.21$	
				$\log K_3 = 16.76 \pm 0.15$	
TLA	tert-Butylbenzene	HNO$_3$	1 (Na,H)X	$\log K_1 = 4.04 \pm 0.04$	30 + 25
				$\log K_2 = 9.97 \pm 0.22$	
				$\log K_3 = 16.04 \pm 0.18$	
TLA	tert-Butylbenzene	HNO$_3$	5 (Na,H)X	$\log K_1 = 4.235 \pm 0.023$	25
				$\log K_2 = 10.43 \pm 0.21$	
				$\log K_3 = 17.00 \pm 0.17$	
TLA	n-Octane	HNO$_3$	1 (Na,H)X	$\log K_1 = 2.51 \pm 0.04$	30 + 20
				$\log K_{2,1} = 4.34 \pm 0.14$	
				$[\log K_8 = 39.98 \pm 0.13]$	
				$[\log K_{30} = 155.27(+ 0.46)]*$	
TLA	n-Dodecane	HNO$_3$	1 (Na,H)X	$\log K_1 = 2.57 \pm 0.15$	30 + 25
				$\log K_{2,1} = 4.61 \pm 0.13$	
				$[\log K_8 = 40.85 \pm 0.13]$	
				$[\log K_{40} = 211.13(+ 0.53)]*$	
DINA	Chloroform	HCl	1 (Li,H)X	$\log K_2 = 15.479 \pm 0.014$	33
TLA	Chloroform	HCl	1 (Li,H)X	$\log K_1 = 6.48 \pm 0.03$	34
THA	Benzene	HCl	1 (Li,H)X	$\log K_1 = 3.73 \pm 0.05$	35
				$\log K_2 = 9.49 \pm 0.05$	
TOA	Benzene	HCl	1 (Li,H)X	$\log K_1 = 3.69 \pm 0.05$	36
				$\log K_2 = 9.30 \pm 0.05$	
TLA	Benzene	HCl	1 (Li,H)X	$\log K_1 = 3.78 \pm 0.06$	37
				$\log K_2 = 9.21 \pm 0.02$	
TLA	o-Xylene	HCl	1 (Li,H)X	$\log K_1 = 3.29 \pm 0.02$	38 + 39
				$\log K_3 = 13.05 \pm 0.04$	
TLA	n-Hexane	HCl	1 (Li,H)X	$\log K_1 = 0.97 \pm 0.03$	40
				$\log K_3 = 9.34 \pm 0.03$	
				(precipitation)	
THA	n-Octane	HCl	1 (Li,H)X	$\log K_1 = 2.21 \pm 0.20$	41
				$\log K_3 = 11.93 \pm 0.08$	
TIOA	Benzene	HCl	1 —	$\log K_1 = 3.81 \pm 0.02$	23
				$\log K_2 = 8.83 \pm 0.04$	
TIOA	Xylene	HCl	1 —	$\log K_1 = 3.35 \pm 0.04$	23
				$\log K_2 = 8.32 \pm 0.04$	
TIOA	Carbon tetrachloride	HCl	1 —	$\log K_1 = 3.23 \pm 0.04$	23
				$\log K_2 = 8.26 \pm 0.03$	
TLA	o-Xylene	HBr	1 (Li,H)X	$\log K_1 = 4.20 \pm 0.03$	42
				$\log K_2 = 9.89 \pm 0.20$	
				$\log K_3 = 15.73 \pm 0.20$	
TLA	o-Xylene	HI	1 (Li,H)X	$\log K_1 = 5.59 \pm 0.02$	43
				$\log K_3 = 20.32 \pm 0.05$	
TLA	Chloroform	HClO$_4$	3 (Na,H)X	$\log K_1 = 7.35 \pm 0.04$	44
				$\log K_2 = 17.06 \pm 0.05$	

continued next page

Table **4.8** *continued from previous page*

Amine	Diluent	Acid	Ionic Medium	$\log K_n \pm 3\sigma$ ($\log K_n$)	Ref.
TOA	o-Xylene	$HClO_4$	1 (Li,H)X	$\log K_1 = 5.74 \pm 0.08$ $\log K_2 = 14.12 \pm 0.08$ $[\log K_8 = 62.07(+0.24)]*$	45
TLA	o-Xylene	$HClO_4$	1 (Na,H)X	$\log K_1 = 5.46 \pm 0.09$ $\log K_2 = 13.88 \pm 0.06$ $[\log K_8 = 59.89(+0.21)]*$	46
TLA	n-Dodecane	$HClO_4$	1 (Li,H)X	$\log K_1 = 2.34(+0.44)$ $\log K_{2,1} = 5.59 \pm 0.03$ $\log K_2 = 11.25 \pm 0.01$ $[\log K_8 = 54.12 \pm 0.15]$	46

THA = tri-n-hexylamine
TOA = tri-n-octylamine
TLA = tri-n-dodecylamine = trilaurylamine
TIOA = tri-iso-octylamine
DINA = di-iso-nonylamine
*Here $\log (K + 3\sigma(K))$ is given within parentheses

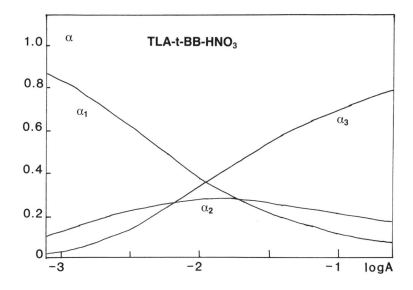

Fig. 4.21 — The fraction of amine present as monomer, α_1, dimer, α_2, and trimer, α_3, against $\log A$ for $Z = 1$ and the system TLA–*t*-BB–HNO_3.

Table 4.9 — Equilibrium constants computed for reaction (4.35) from the constants in Table 4.8

Amine salt	Diluent	$\log k_n$
THANO$_3$	Benzene	$\log k_2 = 1.63$
TLAHNO$_3$	Toluene	$\log k_2 = 1.64$; $\log k_3 = 3.13$
TLAHNO$_3$	m-Xylene	$\log k_2 = 1.70$; $\log k_3 = 3.57$
TOAHNO$_3$	o-Xylene	$\log k_2 = 1.51$; $\log k_3 = 3.36$
TLAHNO$_3$	o-Xylene	$\log k_2 = 1.46$; $\log k_3 = 3.26$
TLAHNO$_3$		$\log k_2 \equiv 1.89$; $\log k_3 \equiv 3.92$ (I = 1)
	$tert$-Butylbenzene	$\log k_2 = 1.96$; $\log k_3 = 4.30$ (I = 5)
TLAHNO$_3$	n-Octane	$[\log k_8 = 19.9$; $\log k_{30} \approx 80]$
TLAHNO$_3$	n-Dodecane	$[\log k_8 = 20.3$; $\log k_{40} \approx 108]$
THAHCl	Benzene	$\log k_2 = 2.03$
TOAHCl	Benzene	$\log k_2 = 1.92$
TLAHCl	Benzene	$\log k_2 = 1.65$
TLAHCl	o-Xylene	$\log k_3 = 3.18$
TLAHCl	n-Hexane	$\log k_3 = 6.43$
THAHCL	n-Octane	$\log k_3 = 5.30$
TIOAHCl	Benzene	$\log k_2 = 1.21$
TIOAHCl	Xylene	$\log k_2 = 1.62$
TIOHCl	Carbon tetrachloride	$\log k_2 = 1.80$
TLAHBr	o-Xylene	$\log k_2 = 1.49$; $\log k_3 = 3.1$
TLAHI	o-Xylene	$\log k_3 = 3.55$
TLAHClO$_4$	Chloroform	$\log k_2 = 2.4$
TOAHClO$_4$	o-Xylene	$\log k_2 = 2.64$; $[\log k_8 = 16.2]$
TLAHClO$_4$	o-Xylene	$\log k_2 = 2.96$; $[\log k_8 = 16.2]$
TLAHClO$_4$	n-Dodecane	$\log k_2 = 6.57$; $[\log k_8 = 35.4]$

To demonstrate that formation of dimers is sufficient, K_2 has been computed for each experimental point as follows. The mass-balance conditions for acid and amine give

$$AZ = 2K_2u^2 \qquad A = a + 2K_2u^2 \qquad (u = ah)$$

From data $Z(\log h)_A$ K_2 has been computed and the average obtained for each amine concentration. The results are given in Table 4.10. There is no trend in K_2 with

Table 4.10 — The system: DINA–CHCl$_3$–HCl. $T = 298$ K. Data from [33].

A, mM	$K_2 \times 10^{-15}$	δZ
6.1	2.49 ± 0.44	-0.0105
14.9	2.28 ± 0.51	-0.0201
21.6	3.68 ± 0.53	$+0.0330$
29.4	3.01 ± 0.65	$+0.0064$
58.7	2.61 ± 0.78	$+0.0003$
151.8	2.96 ± 0.76	$+0.0089$
295.0	2.88 ± 0.64	$+0.0102$
Average of all data	2.83 ± 0.74	

increasing A, showing that further aggregation is negligible. Further experimental evidence in favour of dimers only is given in the original publication. Using the average K_2 obtained from all the experimental points, Z has been computed for each point and U and $\sigma(Z)$ computed from Eqs. (4.32a,b). The results are given in Table 4.11. Using the average K_2 the difference $\delta Z = Z_{calc} - Z_{exp}$ has been computed and the average taken for each A. These values are also given in Table 4.10. Finally U and $\sigma(Z)$ have been computed and given in Table 4.11, the values for K_2, U and $\sigma(Z)$

Table 4.11 — System: DINA–CHCl$_3$–HCl. $T = 298$ K. Data from [33].

Method	K_2	δZ	U	$\sigma(Z)$
K_2 computed for each exp point	$(2.83 \pm 0.74) \times 10^{15}$	0	0.0571	± 0.0226
K_2 computed for each exp point	$(2.83 \pm 0.74) \times 10^{15}$	$\neq 0$	0.0287	± 0.0160
LETAGROP ZETA	$(3.01 \pm 0.03) \times 10^{15}$	$\neq 0$	0.0229	± 0.0143

obtained by using LETAGROP ZETA with $\delta Z \neq 0$ found originally [33]. By introducing δZ the fit is greatly improved. The best fit is obtained with LETAGROP because here K_2 and δZ are varied simultaneously.

These results indicate that the tendency to formation of large aggregates might be real, although the exact nature of the species formed cannot be obtained for the reasons mentioned above.

From Tables 4.8 and 4.9 a number of conclusions can be drawn:

1: Aggregation is more extensive in aliphatic than in aromatic diluents.
2: The shorter the chain, the larger the aggregates formed.
3: Branching of the chain lowers aggregation.
4: With halides, formation of trimers predominates over dimers. Benzene is the only diluent where dimers are formed instead of trimers.

These findings are as expected from the influence of steric effects and solvating power of the diluent. There is a striking confirmation in the isolation of a solid 1:1 complex between tetra-n-butylammonium nitrate and benzene [30]. In aliphatic diluents there seems to be a tendency towards self-solvation by the formation of $(TLA)_2HNO_3$ and $(TLS)_2HClO_4$ in octane and dodecane.

The ability of chloroform to prevent aggregation is most likely due to hydrogen-bond formation to the amine, as has been found for n-octanol, cf. below.

Generally, aggregation increases with decreasing hydrogen-bond strength. If the charge on the NH^+ proton is not neutralized by hydrogen-bond formation, the proton seeks relief in solvation, and if this is not satisfactory, through aggregation.

4.2.2   Formation of mixed complexes

Formation of mixed complexes is a rule rather than an exception in extraction. Sometimes mixed extractants perform better than the separate components. This is used to advantage in synergic extraction. Here, two cases will be discussed, the role of modifiers and the formation of mixed acid–water complexes.

4.2.2.1 Mixed amine–alcohol complexes
In practical extraction, a second organic phase is often formed, from which most of the diluent is squeezed out. This important practical difficulty has been overcome by addition of a modifier, often a long-chain alcohol.

In order to obtain information about speciation in such systems a systematic study was undertaken by Muhammed [48–50] using freezing-point measurements, calorimetry and two-phase titrations to study possible complex formation between trilaurylamine and n-octanol in benzene.

The freezing-point measurements revealed formation of dimers and tetramers of the alcohol. Moreover, several alcohol–water complexes were found. The data were treated both graphically and numerically by the version SUMPA [51]: the results are given in Table 4.12. The extent of formation of hydrates is small compared to

Table 4.12 — Aggregation of n-octanol in dry and wet benzene. $T \approx 279$ K. Data from [48]. Concentration scale is molality = mol/kg

Diluent	Reaction	$\log K$
C_6H_6 "dry"	$2ROH \leftrightharpoons (ROH)_2$	-0.39 ± 0.11
	$4ROH \leftrightharpoons (ROH)_4$	1.36 ± 0.02
C_6H_6 "wet"	$2ROH + H_2O \rightleftharpoons (ROH)_2H_2O$	$2.03(+0.43)$
	$4ROH + 2H_2O \rightleftharpoons (ROH)_4(H_2O)_2$	$4.63(+0.79)$
	$[6ROH + 3H_3O \rightleftharpoons (ROH)_6(H_2O)_3$	$8.29(+0.80)]$

formation of unhydrated species. With knowledge of the octanol equilibria, the reaction between alcohol and amine was studied both by freezing point and calorimetry. The freezing-point measurements indicated formation of a 1:1 complex between amine and alcohol. The calorimetric measurements gave $\triangle H$ for the same process, permitting computation of the equilibrium constant at 298 K. The results are given in Table 4.13.

Table 4.13 — The equilibrium constant for the reaction: $TLA + ROH \leftrightharpoons TLAROH$ (C_6H_6). Data from [49]

Temperature (K)	$\triangle H$ kJ/mole	K kg/mole	K dm³/mole
~279	-14.8	3.64	—
298	-14.8	2.40	2.75

With knowledge of these equilibria and the aggregation equilibria in the system $TLA–C_6H_6–HCl$ [20], the extraction of HCl by TLA in alcohol–benzene mixtures was studied by using two-phase titrations [50]. The data could be described by the following reaction(s):

$$TLA(org) + H^+ + Cl^- + ROH(org) \rightleftharpoons TLAROHHCl(org)$$
$$\log K = 5.41 \pm 0.07$$

$$[2TLA(org) + 2H^+ + 2Cl^- + 2ROH(org) \rightleftharpoons (TLAROH\ HCl)_2\ (org)]$$
$$\log K = 12.0\ (\pm 0.2)$$

Formation of the dimeric species is uncertain, so this species is put in brackets. In Fig. 4.22, the fraction of amine present in various species (α_i) is plotted against the total

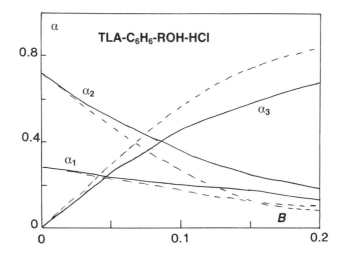

Fig. 4.22 — The fraction of amine present as monomer, α_1, dimer, α_2, and as alcohol–amine hydrochloride complex, α_3. The dashed curves refer to presence of the dimeric mixed complex. Here α_3 is the sum of the two mixed species. $Z = 1$, $[TLAHCl] = 100$ mM, $B = $ total alcohol concentration.

concentration of alcohol, B, when increasing amounts of alcohol are added to a 100 mM solution of TLA in benzene at $Z = 1$. The dashed curves are those obtained by considering the dimeric species $(TLAROHHCl)_2$. Besides the mixed species, only TLAHCl and $(TLAHCl)_2$ need to be taken into account. From Fig. 4.22, it is evident that addition of alcohol decreases the aggregation. If the formation of the second organic phase is related to increased aggregation, the role of the alcohol is simply to prevent further aggregation by forming mixed complexes.

This study shows the importance of building up a picture of complicated systems from simple ones by studying the equilibria in each binary system before study of the complicated mixture.

4.2.2.2 The system TLA–toluene –H₂SO₄ –H₂O

4.2.2.2   The system TLA–toluene –H_2SO_4 –H_2O
As an example of the formation of mixed acid–water complexes the extraction of sulphuric acid by TLA in toluene is considered [35–36]. In Fig. 4.23, Z is plotted

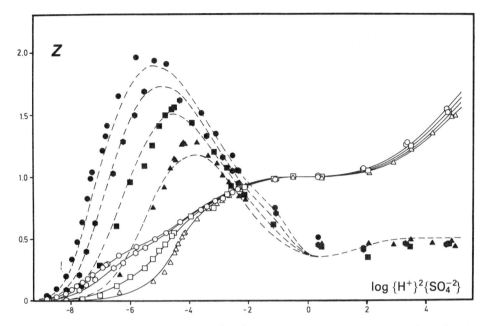

Fig. 4.23 — Z plotted against $\log\{H^+\}^2\{SO_4^{2-}\}$ for the system TLA–toluene–H_2SO_4–H_2O.

$----Z_{H_2O}$ $-----Z_{H_2SO_4}$

△ ▲ $A = 10$ mM □ ■ $A = 41.3$ mM
○ ● $A = 100$ mM ○ ● $A = 157.2$ mM

The curves were computed with the constants given in Table 4.14.

against $\log\{H^+\}^2\{SO_4^{2-}\}$. The dashed curves refer to Z_{H_2O} and the full-drawn curves to $Z_{H_2SO_4}$. The reaction for the extraction of water and acid can be written

$$p\text{TLA(org)} + 2q\text{H}^+ + q\text{SO}_4^{2-} + r\text{H}_2\text{O}$$
$$\rightleftharpoons (\text{TLA})_p(\text{H}_2\text{SO}_4)_q(\text{H}_2\text{O})_r(\text{org}) \qquad (K_{pqr})$$

Table 4.14 gives the species and constants found. First, various graphical methods

Table 4.14 — The constants K_{pqr} and species found to describe the system: TLA–toluene–H_2SO_4–H_2O. $T = 298$ K. Data from [52]

Species	$\log K_{pqr}$
$(\text{TLA})_2(\text{H}_2\text{SO}_4)(\text{H}_2\text{O})_2 \equiv (\text{TLAH}^+)_2\text{SO}_4^{2-}(\text{H}_2\text{O})_2$	$\log K_{212} = 7.52 \pm 0.22$
$(\text{TLA})_4(\text{H}_2\text{SO}_4)_2(\text{H}_2\text{O})_{10} \equiv (\text{TLAH}^+)_4(\text{SO}_4^{2-})_2(\text{H}_2\text{O})_{10}$	$\log K_{4210} = 16.89 \pm 0.19$
$(\text{TLA})_5(\text{H}_2\text{SO}_4)_4(\text{H}_2\text{O})_6$	$\log K_{546} = 27.88 \pm 0.20$
$(\text{TLA})_3(\text{H}_2\text{SO}_4)_3(\text{H}_2\text{O})$	$\log K_{331} = 17.40 \pm 0.05$
$(\text{TLA})_2(\text{H}_2\text{SO}_4)_3(\text{H}_2\text{O})$	$\log K_{231} = 7.43 \pm 0.17$

were used to get information about possible species. The final computations were done with version ZEHTA of LETAGROP. Here many different combinations of complexes were tried. The model in Table 4.14 is a minimum description. The curves in Fig. 4.23 were computed from the model in Table 4.14. In Fig. 4.24, the fraction of amine, α_{pqr}, is plotted against $\log\{H^+\}^2\{SO_4^{2-}\}$ for $A = 49.3$ mM. The species 212 was confirmed by an emf study at unit ionic strength [36]. This investigation shows the importance of also measuring the co-extraction of water if a complete picture of the extraction process is desired.

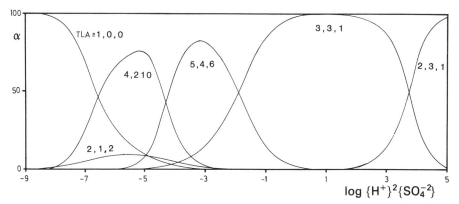

Fig. 4.24 — The fraction of amine, α_{pqr}, present in each species plotted against $\log\{H^+\}^2\{SO_4^{2-}\}$ for $A = 49.3$ mM.

4.2.3 Aggregation of amine salts studied by VPO

4.2.3.1 The method
Salts of long-chain amines can often be crystallized and purified by recrystallization. Their solutions in dry diluents can be studied by vapour-phase osmometry (VPO). This method is similar to cryoscopy and ebullioscopy, in that a temperature difference $\triangle T$ measured between pure solvent and one containing solute is proportional to the sum of all dissolved entities in dilute solutions.

In practice, the resistance difference, $\triangle R$, between two matched thermistors is measured in a Wheatstone bridge. Thus

$$\triangle R = k \cdot \sum m_i \quad (m = \text{molality, mol/kg}) \tag{4.36}$$

k is an empirical constant obtained by using a calibrating substance assumed to behave ideally in the concentration range studied. This method has been used to get information about the aggregation of di-isononylammoniumchloride (DINAHCl) in benzene.

4.2.3.2 The system DINAHCl–C₆H₆ at 338 K
The aggregation of DINAHCl in benzene at various temperatures has been studied by the VPO method [54]. Here the data at 65°C (338 K) will be used to illustrate the treatment of such data by graphical methods.

Consider a compound B that aggregates according to reaction (4.35) with formation of several species. The mass-balance condition gives

$$B = b + 2k_2b^2 + 3k_3b^3 + \ldots = \sum_0^n nk_nb^n$$

This process corresponds to set 1 in Fig. 4.12, i.e. association of one component. From the VPO measurements the sum of all species is obtained.

$$S = b + k_2b^2 + k_3b^3 + \ldots = \sum_0^n k_nb^n$$

B is known from the preparation of the solutions of B in the solvent. With knowledge of S and B, the average degree of aggregation, $\bar{n}$, is obtained.

$$\bar{n} = \frac{B}{S} = \sum nk_nb^n / \sum k_nb^n = \frac{\mathrm{dlog}\, S}{\mathrm{dlog}\, b}$$

and the concentration of monomer can be computed from

$$\log b - \log b_0 = \int_{S_0}^{S} (1/\bar{n})\mathrm{dlog}\, S = I \qquad (4.37)$$

where S_0 is the lowest value measured. If aggregation can be neglected at that point, $S_0 = b_0 = B_0$ and the integration constant becomes $\log B_0$.

If some species B_p predominates at the first experimental point, the value obtained for b_0 is:

$$b_0 = \frac{pS_0 - B_0}{p - 1}$$

The use of this expression requires knowledge of p, which often can be inferred from the nature of the system or from the curve $\bar{n}(B)$.

From Eq. (4.37)

$$\log b = \log b_0 + I$$

The integral I can be obtained by graphical or numerical integration of the curve $(1/\bar{n})\log S$. From the expression above, it is seen that a plot of $\bar{n}$ against I has the same shape as the plot $\bar{n}(\log b)$.

In Fig. 4.25 $\bar{n}$ is plotted against I for the system DINAHCl–C_6H_6 at 65°C. The data cluster around $\bar{n} \approx 2.2$. It is reasonable to start to consider formation of both dimers and trimers. This gives the following expression for $\bar{n}$

$$\bar{n} = \frac{b + 2k_2b^2 + 3k_3b^3}{b + k_2b^2 + k_3b^3} = \frac{1 + 2k_2b + 3k_3b^2}{1 + k_2b + k_3b^2} = \frac{1 + 2u + 3pu^2}{1 + u + pu^2} \qquad (4.38)$$

where

$$u = k_2b \qquad p = k_3/k_2^2 \qquad\qquad\qquad (4.39a,b)$$

By comparing the curves in Fig. 4.25 with a set of normalized curves on transparent paper the following value was obtained for p:

$$p = 0.05 \pm 0.01$$

With p known, the u-value was computed for each experimental point from Eq. (4.38). With u known, b was obtained from

$$B = b(1 + 2u + 0.15u^2)$$

With u and b known, k_2 and k_3 were computed from Eqs. (39a,b) giving

$$k_2 = (2.9 \pm 0.5)10^3 \qquad k_3 = (4.3 \pm 1.3)10^5$$

With knowledge of k_2 and k_3, consistent values for b were computed from

$$B = b + 2k_2b^2 + 3k_3b^3$$

By using the b-values obtained in this way, the sum S was then computed from

$$S = b + k_2b^2 + k_3b^3$$

From Eq. (4.36) $\triangle R$ is computed as $\triangle R = kS$, where the calibration constant k has been determined by separate experiments. In Table 4.15, experimental and computed $\triangle R$-values are compared. The agreement is satisfactory with the exception of the two highest concentrations. It seems that the VPO method as applied here is satisfactory up to about 0.1 m solutions.

In Fig. 4.26 the fraction of DINAHCl present in various species is plotted vs. the total molality $(m_0 = B)$.

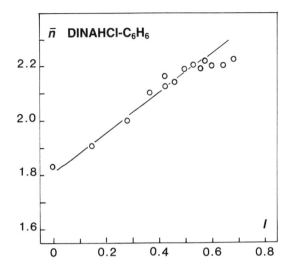

Fig. 4.25 — $\bar{n}$ plotted against I for the system DINAHCl–C_6H_6 at 338 K. The curve was computed with the constants: $k_2 = 2.9 \times 10^3$, $k_3 = 4.3 \times 10^5$.

Table 4.15 — Comparison between computed and experimental $\triangle R$-values for VPO measurements on the systems: DINAHCl–C_6H_6 at 65°C (338.2 K). Concentration scale mol/kg

B, m	$\triangle R$ calc	$\triangle R$ exp	$\dfrac{\text{Diff} \times 100}{\triangle R_{\text{exp}}}$
0.005	1.16	1.14	+ 1.75
0.0115	2.48	2.50	− 0.80
0.0221	4.51	4.54	− 0.66
0.0347	6.84	6.77	+ 1.03
0.0465	8.94	8.92	+ 0.22
0.0475	9.12	8.95	+ 1.90
0.0579	10.93	10.96	− 0.27
0.0700	12.99	12.93	+ 0.46
0.0824	15.07	15.03	+ 0.27
0.0932	16.84	17.03	− 1.12
0.1076	19.16	19.29	− 0.67
0.1178	20.78	21.22	− 2.07
0.1381	23.95	24.65	− 2.84
0.1483	25.50	26.35	− 3.23

4.2.3.3 Systems with only one aggregate

When the formation of only one aggregate gives a satisfactory description the following approach can be used. For reaction (4.35) with formation of B_n only, the mass-balance conditions give

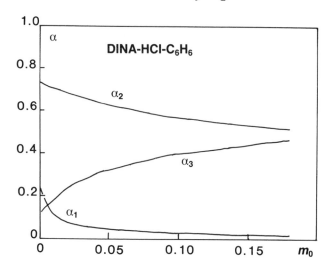

Fig. 4.26 — The fraction of DINAHCl present in monomer, α_1, dimer, α_2, and trimer, α_3, plotted against the total molality $B\ (=m_0)$.

$$B = b + nk_nb^n \qquad S = b + k_nb^n$$

$$\bar{n} = \frac{B}{S} = \frac{b + nk_nb^n}{b + k_nb^n} = \frac{1 + nu}{1 + u} \qquad u = k_nb^{n-1}$$

and

$$u = \frac{\bar{n} - 1}{n - \bar{n}} \qquad k_n = u/b^{n-1} \qquad\qquad\qquad (4.40a,b)$$

Jedináková [55a,b] made VPO measurements on a number of TOA salts in benzene at 25°C and 50°C. Her results have been recalculated using TOA in benzene as standard (to give k in Eq. (4.36)) and assuming formation of either dimers or trimers. The results are given in Table 4.16. As expected, aggregation decreases with temperature. TOADCl and TOADClO$_4$ seem to form stronger aggregates than the corresponding protonated salts. Otherwise there is little difference between protonated and deuterated compounds.

4.2.4 The extraction of HBr and H$_2$O by TLAHBr in o-xylene
Lodhi [42] studied the extraction of water and hydrobromic acid by trilaurylamine dissolved in o-xylene. The potentiometric titrations could be explained by formation of dimers and trimers, with a tendency towards formation of larger aggregates, represented by a 30:30 complex. Nevertheless, the extraction of water and acid in the range of excess of acid ($Z_{HBr} > 1$) can be described by a mechanism independent of

Table 4.16 — Aggregation constants at 298 K and 348 K for TOA-salts in benzene. The constants have been computed from Eqs. (40a,b). Data from [54]

Salt	$\log k_n$ (298 K)	$\log k_n$ (348 K)
TOAHCl	$\log k_2 = 1.62 \pm 0.15$	$\log k_2 = 1.51 \pm 0.19$
TOADCl	$\log k_2 = 2.11 \pm 0.21$	$\log k_2 = 1.56 \pm 0.17$
TOAHClH$_2$O	$\log k_2 = 1.86 \pm 0.14$	$\log k_2 = 1.30 \pm 0.12$
TOADCl.D$_2$O	$\log k_2 = 1.58 \pm 0.18$	$\log k_2 = 1.46 \pm 0.06$
TOA(HCl)$_2$	$\log k_2 = 2.25 \pm 0.15$	$\log k_2 = 1.72 \pm 0.16$
TOAHBr	$\log k_2 = 1.79 \pm 0.12$	$\log k_2 = 1.84 \pm 0.12$
TOAHI	$\log k_3 = 3.08 \pm 0.18$	$\log k_3 = 2.86 \pm 0.18$
TOAHNO$_3$	$\log k_3 = 3.48 \pm 0.06$	$\log k_3 = 3.34 \pm 0.14$
TOADNO$_3$	$\log k_3 = 2.81 \pm 0.09$	$\log k_3 = 2.72 \pm 0.10$
TOAHClO$_4$	$\log k_3 = 3.53 \pm 0.19$	$\log k_3 = 3.19 \pm 0.12$
TOADClO$_4$	$\log k_3 = 3.90 \pm 0.16$	$\log k_3 = 3.50 \pm 0.06$

total amine concentration. In Figs. 4.27 and 4.28 Z_{HBr} and $[H_2O]_{org}$ are plotted vs. the equilibrium molarity of HBr in the aqueous phase. From Fig. 4.27 it is evident that excess of acid starts to extract at about $5\,M$ HBr. Within the experimental uncertainty, all the data points fall on the same curve. The same applies to H_2O. The data were fitted with the following reaction

$$\text{TLAHBr(org)} + H^+ + Br^- + nH_2O \rightleftharpoons \text{TLA (HBr)}_2(H_2O)_n \text{ (org) } (K_n')$$

with

$$K_3' = 1.2 \times 10^{-4} \qquad K_1' = 5.1 \times 10^{-6}$$

Before the mixed acid–water complexes start to extract, water decreases in the organic phase according to the reaction

$$H_2O(aq) \rightleftharpoons H_2O(\text{TLAHBr}) \qquad K = 0.593[\text{TLAHBr}]$$

i.e. there is a simple distribution of water between the aqueous phase and the amine salt. This reaction was studied by shaking samples of TLAHBr in o-xylene with LiBr–H$_2$O mixtures of known water activity.

The acid extracted by the diluent was the subject of a separate study, and the water extracted by the diluent could be corrected by using data from the literature [6].

4.2.5 The solubility problem
It is of great practical importance that the amine and its salts should not be too soluble in the aqueous phase, to prevent losses from becoming prohibitively expensive. The solubilities of long-chain amines are so small that direct determination might be difficult. Therefore a combination of experiment and empirical equations will be used.

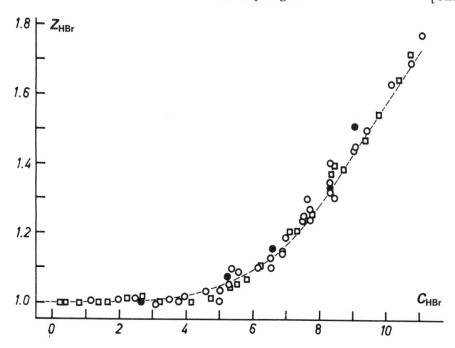

Fig. 4.27 — Z_{HBr} plotted against the equilibrium molarity of HBr, C_{HBr}, for the system
TLA–*o*-xylene–HBr.

$\square\ A = 76.2\ \text{m}M$ ● $A = 231.6\ \text{m}M$ ○ $A = 307.0\ \text{m}M$

The curve was computed from the constants given in the text. Reprinted from D. Dyrssen, J.-
O. Liljenzin and J. Rydberg, editors, *Solvent Extraction Chemistry*, 1967, p. 423, by permission
of the copyright holders, Elsevier Publishing Co.

The distribution coefficients of amine and its salt AHX are needed

$$\lambda_A = [A]_{org}/[A]_{aq} \qquad \lambda_{AHX} = [AHX]_{org}/[AHX]_{aq}$$

Dyrssen [56] estimated λ_A and λ_{AHX} for nitrates and chlorides of tertiary ammonium
salts. For primary, secondary and tertiary amines he found for aromatic diluents

$$\log \lambda_A = 0.6n' - 2.3$$

where n' is the number of carbon atoms in the amine. The equilibrium constant K_1 of
reaction (4.25) is related to λ_A and λ_{AHX} by

$$\log K_1 = \log \lambda_{AHX} - \log \lambda_A + pK_a$$

where K_a is the acid dissociation constant of AHX in the aqueous phase. Dyrssen
took $pK_a = 10.6$ as a suitable average. In order to get an idea of the distribution
coefficients in aliphatic diluents, $\lambda_{TOAHCl} = 170$ in heptane and 8 M HCl [57] and

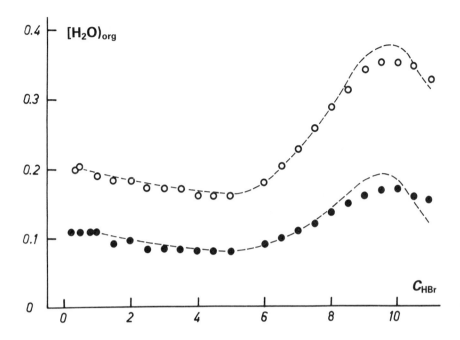

Fig. 4.28 — $[H_2O]_{org} - [H_2O]_{dil}$ plotted against C_{HBr} for the system TLA–o-Xylene–HBr.
● $A = 321.6\,mM$ ○ $A = 307.0\,\text{mM}$
The curves were computed from the constants given in the text. Reprinted from D. Dyrssen, J.-O. Liljenzin and J. Rydberg, editors, *Solvent Extraction Chemistry*, 1967, p. 423, by permission of the copyright holders, Elsevier Publishing Co.

$\log K_1 \approx 1.0$ for TLAHCl in n-hexane [40] have been used to estimate distribution coefficients in aliphatic diluents. Table 4.17 gives the expressions arrived at. In Table

Table 4.17 — Expressions for λ_A and λ_{AHX} in aliphatic and aromatic diluents. $T = 298$ K

Acid	Expression	Diluent
—	$\log \lambda_A = 0.6n' - 2.3$	Aromatics
—	$\log \lambda_A = 0.6n' - 2.6$	Aliphatics
HNO_3	$\log \lambda_{AHNO_3} = 0.6n' - 8.4$	Aromatics
HNO_3	$\log \lambda_{AHNO_3} = 0.6n' - 10.7$	Aliphatics
HCl	$\log \lambda_{AHCl} = 0.6n' - 9.3$	Aromatics
HCl	$\log \lambda_{AHCl} = 0.6n' - 12.2$	Aliphatics
HBr	$\log \lambda_{AHBr} = 0.6n' - 8.7$	Aromatics
HI	$\log \lambda_{AHI} = 0.6n' - 7.3$	Aromatics
$HClO_4$	$\log \lambda_{AHClO_4} = 0.6n' - 7.4$	Aromatics
$HClO_4$	$\log \lambda_{AHClO_4} = 0.6n' - 10.7$	Aliphatics

Table 4.18 — Estimated distribution coefficients for some tertiary amines and their salts. Concentration unit $M(= \text{mol dm}^{-3})$. $T = 298$ K

R_3N	n'	$\log \lambda_A$	$\log \lambda_{AHNO_3}$	$\log \lambda_{AHCl}$	$\log \lambda_{AHBr}$	$\log \lambda_{AHI}$	$\log \lambda_{AHClO_4}$
Triethylamine	6	1.3	− 4.8	− 5.7	− 5.1	− 3.7	− 3.8
Tributylamine	12	4.9	− 1.2	− 2.1	− 1.5	− 0.1	− 0.2
Trihexylamine	18	8.5	2.4	1.5	2.1	3.5	3.4
Trioctylamine	24	12.1	6.0	5.1	5.7	7.1	7.0
Trilaurylamine	36	19.3	13.2	12.3	12.9	14.3	14.2

4.18 some estimated distribution coefficients are given for a number of tertiary amines. According to this table TOA is the lower limit for negligible solubility of the amine salts. Nevertheless our studies with tri-n-hexylamine (THA) have revealed no solubility problems, cf. Table 4.8.

4.2.6 Metal extraction
Our group has studied the extraction of several metals. One question of interest has been the way in which aggregation influences the metal extraction. Some typical results are discussed.

4.2.6.1 The extraction of iron(III) by TLAHCl
Kuča [58] studied the extraction of iron(III) by TLAHCl dissolved in o-xylene. From the work of Tavares $et\ al.$ [38, 39], the formation of monomer and trimer are well established, together with something large, formally represented by a 50-mer. However, the best fit was obtained by neglecting aggregation, as was found for the extraction of water and HBr by TLAHBr in o-xylene.

In Fig. 4.29 Z is plotted against $\log [\text{Fe(III)}]_{aq,tot}$. As usual, Z is the number of iron ions per amine, i.e., $Z = [\text{Fe(III)}]\text{org,tot}/A$. For the reaction

$$q\text{FeCl}_3(\text{aq}) + p\text{TLAHCl}(\text{org}) \rightleftharpoons (\text{TLAHCl})_p(\text{FeCl}_3)_q(\text{org}) \quad (K_{pq})$$
$$(4.41)$$

the best description was obtained by neglecting aggregation of the amine salt. Table 4.19 gives the constants and complexes obtained. Besides the model accepted (set 1) the best sets obtained when taking aggregation into account are mentioned, set 2 where both trimer and a 48-mer are considered and one, set 3, where only trimer is considered.

If large aggregates are fomed they can be treated as a separate phase, where all sites are equal (set 1). Formation of large aggregates is thus not contradictory to the description used by Kuča. This is well illustrated by the liquid cation-exchanger dinonylnaphthalenesulphonic acid, which can be treated by use of a model that also applies to solid resins, i.e. as a separate phase, cf. [59].

4.2.6.2 The extraction of zinc by TOAHCl
Aguilar [60] studied the extraction of ^{65}Zn by TOAHCl in benzene. He considered the following reaction

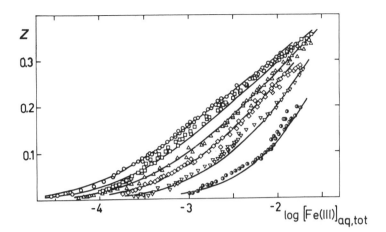

Fig. 4.29 — $Z_{Fe(III)}$ plotted against $\log[FeIII)]_{aq,tot}$ for the system TLAHCl–o-xylene–Fe(III).
The ionic strength $I = 1\,M$ (Li,H)Cl. $T = 298$ K.
  ◑ $A = 10$ mM   ▽ $A = 50$ mM   ◇ $A = 100$ mM
  △ $A = 150$ mM   □ $A = 220$ mM   ○ $A = 300$ mM
The curves were computed from the constants given in Table 4.19.

Table 4.19 — Equilibrium constants for reaction (4.41). $T = 298$ K. Data from [58]. (Reproduced with permission of the copyright holder, John Wiley Inc.)

Set	Species	$\log K_{pq}$	$\sigma(Z)$
1	0,1; 1,1; 2,1; 3,1	$\log K_{11} = 1.03 \pm 0.03$ $\log K_{21} = 2.75 \pm 0.03$ $\log K_{31} = 3.68 \pm 0.05$	± 0.0087
2	0,1; 0,3; 0,48; 4,1; 5,2; 5,3	—	± 0.024
3	0,1; 0,3; 1,1; 3,1; 3,3; 9.3	—	± 0.0094

$$qZnCl_2(aq) + pTOAHCl(org) \rightleftharpoons (TOAHCl)_p(ZnCl_2)q \qquad (K_{pq}) \quad (4.42)$$

The distribution coefficient was found to be independent of the total zinc concentration, i.e. $q = 1$. Aguilar treated the distribution data in the following way. The mass-balance conditions for zinc in the two phases give

$$[Zn]_{aq} = [Zn^{2+}] + \sum \beta_n[Zn^{2+}][Cl^-]^n = [Zn^{2+}]\sum \beta_n$$

since $[Cl^-] = 1$ and $\beta_0 = 1$

The ionic strength in the aqueous phase was kept at $I = 1.00$ by (Li,H)Cl.

$$[Zn]_{org} = \beta_2[Zn^{2+}][Cl^-]^2 \sum K_{p1}[TOAHCl]^p$$

$$= \beta_2[Zn^{2+}] \sum K_{p1}[TOAHCl]^p$$

giving

$$D = \frac{[Zn]org}{[Zn]aq} = \frac{\beta_2}{\Sigma\beta_n} \times \sum K_{p1}a^p = \sum Q_p a^p$$

where

$$a = [TOAHCl] \qquad Q_p = (\beta_2/\Sigma\beta_n)K_{p1}$$

In benzene, TOAHCl forms dimers, cf. Tables 4.10 and 4.16. From these tables, it is seen that $\log k_2 \sim 1.9$. Aguilar used $\log k_2 = 1.92$. From the mass-balance condition for TOAHCl, the concentration of monomer was computed, i.e.

$$[TOAHCl] = A = a + 2 \times 10^{1.92}a^2 + \sum \beta_2 K_{p1}a^p \approx a + 2 \times 10^{1.92}a^2$$

The total zinc concentration is practically negligible compared with the total concentration of TOAHCl. Using a computed from the expression above as ligand variable, Aguilar found that the data could best be described by assuming formation of species with $p = 2$ and $p = 3$. This gives the following expression for D.

$$D = Q_2a^2 + Q_3a^3 = Q_2a^2(1 + (Q_3/Q_2)a)$$

which can easily be normalized by

$$Y = D/Q_2a^2 = (1 + v) \qquad \text{or} \qquad Y' = DQ_3^2/Q_2^3 = v^2(1 + v) \qquad (4.43a,b)$$

with

$$v = (Q_3/Q_2)a \qquad\qquad\qquad\qquad\qquad\qquad\qquad (4.43c)$$

In Fig. 4.30, data for $\log D/a^2$ plotted vs. $\log a$ are compared with normalization (4.43a) in the position of best fit. Aguilar used normalization (4.43b). In Table 4.20

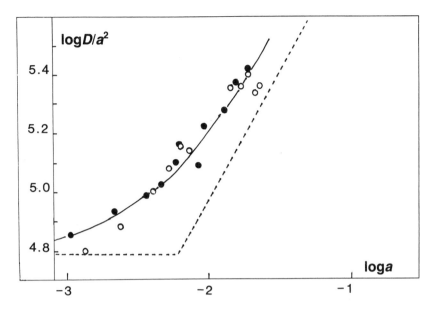

Fig. 4.30 — log D/a^2 plotted against log a and compared with the normalized curve in the position of best fit for the system TOAHCl–C_6H_6–Zn(II). The ionic strength $I = 1\ M$ (Li,H)Cl. $T = 298$ K.

$\bigcirc\ [Zn]_{tot} = 1.2 \times 10^{-5}\ M$ $\bullet\ [Zn]_{tot} = 5 \times 10^{-6}\ M$

the constants obtained in this way are compared with those found by Aguilar both graphically and with LETAGROP DISTR. The agreement is satisfactory.

Table 4.20 — Equilibrium constants for reaction (4.42) obtained in different ways. $T = 298$ K. Data from [60]

Method	log Q_2	log Q_3	U	$\sigma(\log D)$
Normalization log (D/a^2), log a	4.79	7.00	5.04×10^{-2}	± 0.049
Normalization log D, log a	4.80	7.00	5.07×10^{-2}	± 0.049
LETAGROP DISTR	4.82 ± 0.09	6.97 ± 0.08	4.92×10^{-2}	± 0.048
Suggested values	4.80	6.97	5.16×10^{-2}	± 0.050
Normalization log (D/A), log A	log $Q_1' = 1.95$	log $Q_2' = 4.27$	0.410	± 0.140

If instead of a, the monomer concentration, the total concentration A is used, the constants Q_1' and Q_2' are obtained, i.e. $p = 1$ and $p = 2$. The fit is much worse, and in the system studied the aggregation is too small to permit treatment of the extractant as a separate phase.

Aguilar studied the extraction of zinc with THAHCl and TLAHCl in benzene. Table 4.21 gives the constants found for reaction (4.41). The influence of chain length seems to be small and practically negligible.

Table 4.21 — Constants for reaction (4.42) for three tertiary amines in benzene. $T = 298$ K

Amine	$\log Q_2$	$\log Q_3$	Reference
THA	4.68 ± 0.07	7.05 ± 0.04	61
TOA	4.82 ± 0.09	6.97 ± 0.08	60
TLA	4.66 ± 0.02	6.52 ± 0.04	62

4.2.6.3 Species in the organic phase

Spectroscopic studies of the organic phase reveal that most often the complex corresponding to the characteristic co-ordination number is formed in this phase, independent of the ligand activity in the aqueous phase. Table 4.22 contains an excerpt from an earlier review [63].

Table 4.22 — Species identified by spectroscopy in amine systems. From [63]

Amine system	Species
TIOA–toluene	$CuCl_4^{2-}$
TOA–toluene	$MnCl_4^{2-}$
TIOA–toluene	$FeCl_4^{-}$
THA–benzene	$FeCl_4^{-}$
TIOA–toluene	$CoCl_4^{2-}$
THA–benzene	$FeCl_4^{-}$
TOA–toluene	$NiCl_4^{2-}$
TOA–xylene	UCl_6^{2-}
TOA–xylene	$NpCl_6^{2-}$
TOA–xylene	$PuCl_6^{2-}$
TOA–xylene	$UO_2Cl_4^{2-}$
TOA–xylene	$NpO_2Cl_4^{2-}$
TOA–xylene	$PuO_2Cl_4^{2-}$
TIOA–toluene	$CuBr_4^{2-}$
TOA–toluene	$MnBr_4^{2-}$
TIOA–toluene	$FeBr_4^{-}$
TOA–xylene	$Th(NO_3)_6^{2-}$
TOA–xylene	$U(NO_3)_6^{2-}$
TOA–xylene	$Np(NO_3)_6^{2-}$
TOA–xylene	$Pu(NO_3)_6^{2-}$
TOA–xylene	$UO_2(NO_3)_3^{-}$
TOA–xylene	$NpO_2(NO_3)_3^{-}$
TOA–xylene	$PuO_2(NO_3)_3^{-}$

4.2.6.4 Influence of amine structure on selectivity

Sometimes selectivity varies very much with amine structure. This is illustrated in Table 4.23 [64]. Here, the distribution coefficient is given for U(VI) between 1 M sulphate solution at pH = 1 and 0.1M solutions of the amine in various diluents: kerosene, benzene and chloroform. In kerosene, formation of a third phase and insolubility of the amine salt give problems. For these, chloroform is a possible diluent, but otherwise chloroform is poorer than benzene and kerosene, probably

Table 4.23 — Distribution coefficients for U(VI) extraction by various amines. pH = 1, $[SO_4^{2-}] = 1M$, $A = 0.100M$. From [64]

Amine	D		
	Kerosene	Benzene	Chloroform
Primary and secondary amines			
Primene JM-T	3	10	90
Di-n-dodecylamine	†	90	100
Di(tridecyl)amine	80	120	40
Tertiary amines			
Methyl-di-n-decylamine	†	50	50
TOA	30	100	5
Tris(tridecyl)amine	140	60	0.3
Tris(2-ethylhexyl)amine	0.1	0.2	< 0.1
Dialkyl-laurylamines			
Laurylamine	—	—	(1)
Dimethyl-laurylamine	—	< 0.1	7
Diethyl-laurylamine	—	< 0.1	20
Dibutyl-laurylamine	†	95	10
Dihexyl-laurylamine	†	110	6
Alkyldilaurylamines			
Dilaurylamine	†	80	100
Methyldilaurylamine	6	70	40
Butyldilaurylamine	—	140	7
Benzyldilaurylamine	15	50	4

†Third phase formation

because of hydrogen bonding to the amine. Of the various kinds of amines reported, alkyldilaurylamines give rather good distribution coefficients while dialkyl-laurylamines give solubility problems. TOA gives a good distribution coefficient in benzene while tris(decyl)amine gives the best distribution coefficient in kerosene of the tertiary amines. This partly explains the large amount of work in the nuclear field done with TLA and its relatives.

4.2.7 Concluding remarks
The long -chain amines and their salts form an interesting group of compounds with an interesting physical chemistry. One topic of interest is the minimum size of aggregates that will permit them to be treated as a separate phase. A study of metal extraction from amines in aliphatic diluents might give answer to that question.

Acknowledgements
My sincere thanks to all who have participated in the amine project over the years: M. Aguilar, B. Bolander, O. Budevsky, M. Chimboulev, P. R. Danesi, F. Fredlund, I. Hagfeldt, V. Jedinakova, T. Korshunova, L. Kuča, A. Lodhi, J. M. Madariaga, M. Muhammed, S. Poturaj, K. Rasmusson, V. S. Soldatov, J. Szabon, M. Tavares, M. Valiente and B. Warnqvist.

REFERENCES
[1] A. K. De, S. M. Khopkar and R. A. Chalmers, *Solvent Extraction of Metals,* Van Nostrand–Reinhold, London, 1970.
[2] ISEC 83, *International Solvent Extraction Conference,* Proceedings, Denver, Colorado, August 26–September 2, 1983.
[3] D. Naden and M. Streat, eds, *Ion Exchange Technology,* Ellis Horwood, Chichester, 1984.

[4] E. Högfeldt and M. Chimboulev, *Hydrometallurgy*, 1976, **1** 389.
[5] E. Högfeldt and J. Zelinka, *Hydrometallurgy*, 1979, **4**, 337.
[6] E. Högfeldt and K. Rasmusson, *Sv. Kem. Tidskr.*, 1966, **78**, 490.
[7] L. Ödberg and E. Högfeldt, *Acta Chem. Scand*, 1969, **23**, 1330.
[8] E. Högfeldt, *in Metal Complexes in Solution*, E. A. Jenne, E. Rizzarelli, V. Romano and S. Sammartano, eds., Piccin, Padua, 1986, p.1.
[9] M. Gordon, C. Hope, L. Loan and R. Roe, *Proc. Roy. Soc.*, 1960, **A258**, 215.
[10] E. Högfeldt. To be published.
[11] W. F. Giauque, E. W. Hornung, J. E. Kunzler and T. R. Rubin, *J. Am. Chem. Soc.*, 1960, **82**, 62.
[12] E. Högfeldt, F. Fredlund, L. Ödberg and G. Merenyi, Acta Chem. Scand., 1973 **27**, 1781.
[13] D. Dyrssen and T. Sekine, *J. Inorg. Nucl. Chem.*, 1964, **26**, 981.
[14] J. Rydberg, *Acta Chem. Scand.*, 1950, **4**, 1503.
[15] D. Dyrssen and L. G. Sillén, *Acta Chem. Scand.*, 1953, **7**, 663.
[16] J. Bjerrum, *Diss*, Haase, Copenhagen, 1941.
[17] M. H. Mihailov, *J. Inorg. Nucl. Chem.*, 1974, **36**, 107.
[18] D. H. Liem, *Acta Chem. Scand.*, 1971, **25**, 1521.
[19] E. Högfeldt, *Acta Chem. Scand.*, 1952, **6**, 610.
[20] E. Högfeldt, F. Fredlund and K. Rasmusson, *Trans. Royal. Inst. Technol.*, 1964, No. 229.
[21] L. G. Sillén, *Acta Chem. Scand.* 1961, **15**, 1981.
[22] L. G. Sillén and N. Ingri. Personal communication, 1962.
[23] A. Vieux. Personal Communication, 1974.
[24] Y. Marcus, *XI Conference on Coordination Chemistry*, Butterworths, London 1969, p. 85.
[25] E. Högfeldt and F. Fredlund, to be published.
[26] L. Kuča and E. Högfeldt, *Acta Chem. Scand*, 1967, **21**, 1017.
[27] M. Aguilar, *Chem. Scripta*, 1976, **9**, 58.
[28] L. Kuča and E. Högfeldt, *Acta Chem. Scand.*, 1971, **25**, 1261.
[29] E. Högfeldt, F. Fredlund and K. Rasmusson, *Trans. Royal. Inst. Technol.*, 1964, No. 226.
[30] E. Högfeldt, *Proceedings, ICSEC*, Jerusalem 1968, Wiley, New York, 1969. p. 157.
[31] E. Högfeldt, to be published.
[32] E. Högfeldt and F. Fredlund, *Trans. Royal. Inst Technol*, 1964, No. 227.
[33] B. Warnqvist, *Acta Chem. Scand.*, 1967, **21**, 1353.
[34] S. Poturaj and E. Högfeldt, *Acta Chem. Scand.* 1978, **A32**, 85.
[35] M. Aguilar and E. Högfeldt, *Chem. Scripta*, 1973, **3**, 107.
[36] M. Aguilar and E. Högfeldt, *Chem. Scripta*, 1972, **2**, 149.
[37] M. Muhammed, J. Szabon and E. Högfeldt *Chem. Scripta*, 1974, **6**, 61.
[38] L. Kuča, E. Högfeldt and M. Tavares, *Ark. Kemi*, 1971, **32**, 405.
[39] E. Högfeldt and M. de Jesus Tavares, *Trans. Royal Inst. Technol.*, 1964, No. 228.
[40] O. Budevsky, and E. Högfeldt, *Chem. Scripta*, 1974, **5**, 107.
[41] M. Aguilar and M. Valiente, *J. Inorg. Nucl. Chem.*, 1980, **42**, 405.
[42] A. Lodhi, *Ark. Kemi*, 1967, **27**, 309.
[43] E. Högfeldt, to be published.
[44] F. Fredlund and E. Högfeldt, *Chem. Scripta*, 1977, **11**, 217.
[45] E. Högfeldt, to be published.
[46] E. Högfeldt, P. R. Danesi and F. Fredlund, *Acta Chem. Scand.*, 1971, **25**, 1338.
[47] T. J. Plati and E. C. Taylor, *J. Phys. Chem.* 1964, **68**, 3426.
[48] M. Muhammed, *Chem. Scripta* 1975, **8**, 177.
[49] M. Muhammed and R. Arnek, *Chem. Scripta*, 1975, **8**, 187.
[50] M. Muhammed, *Chem. Scripta*, 1975, **8**, 229.
[51] B. Warnqvist, *Chem. Scripta, 1971*, **1**, 49.
[52] E. Högfeldt, M. Madariaga and M. Muhammed, *Acta Chem. Scand.*, 1985, **A39**, 805.
[53] M. Madariaga, M. Muhammed and E. Högfeldt, *Solvent Extraction and Ion Exchange*, 1986, **4**, 1.
[54] E. Högfeldt, to be published.
[55] V. Jedináková and E. Högfeldt, *Chem. Scripta*, 1976, **9**, 171, 178.
[56] D. Dyrssen, *Svensk. Kem. Tidskr.*, 1965, **77**, 387.
[57] W. Smulek and S. Siekierski, *J. Inorg. Nucl. Chem.*, 1962, **24**, 1651.
[58] L. Kuča and E. Högfeldt, *Acta Chem. Scand.*, 1968, **22**, 183.
[59] E. Högfeldt, *Reactive Polymers*, 1984, **2**, 19.
[60] M. Aguilar, *Chem. Scripta*, 1974, **5**, 213.
[61] M. Aguilar, *Chem. Scripta*, 1973, **4**, 207.
[62] M. Aguilar and M. Muhammed, *J. Inorg. Nucl. Chem.*, 1976, **38**, 1193.
[63] E. Högfeldt, in *Ion Exchange, Vol 1*, J. A. Marinsky, ed., Dekker, New York, 1966, p. 139.
[64] E. Strandell, *Report KFT-12*, (Swedish AEC), 1960.

5

Graphical treatment of liquid–liquid equilibrium data

M. Aguilar
Department of Chemistry, ETYSEIB, Polytechnic University of Catalunya, Barcelona, Spain

5.1 INTRODUCTION

Metal extraction reactions have been known since the middle of last century, when several investigators [1,2] studied the application of the extraction of metal-thiocyanate complexes for the determination of metals in rocks. Still in the last century, several other systems were studied and the mathematical law which explains the distribution of neutral species between two inmiscible phases was established [3].

Later, the synthesis of new organic ligands (e.g. diketones, oxine, dithizone) which form neutral metal complexes that are highly soluble in organic solvents, increased and diversified the interest in metal extraction and many new reactions and analytical procedures were studied [4]. The chromophoric properties of most of the new reagents greatly assisted the task, because the extraction of coloured compounds could be followed by spectrophotometric measurements. The work done in this period created the foundations of the use of extraction coupled with photometric determination as a standard method in trace analysis of metals [5].

With the exception of work on organic azo-dyes and some diketones, detailed studies of extraction reactions are scarce, mainly becuase most interest was focused on the analytical applications of the different reactions [4–6].

The application of metal extraction reactions in the nuclear industry represented an important step in the development of solvent extraction chemistry [7]. The introduction of neutral organophosphorus compounds (TBP, TOPO, etc) was closely connected with the search for new reagents for nuclear fuel reprocessing, initiated by the pioneering work at Oak Ridge National Laboratory. The observation that the acid hydrolysis products of the reagent TBP, the mono- and di-butylphosphoric acids, were themselves effective metal extractants, was a stimulus for the use of phosphoric acid derivatives for processing spent reactor fuel. Also metal extraction reactions with long-chain amines interested many workers after it was shown

[7,8] that tertiary amines were useful for recovering active and fissionable metals from irradiated fuel elements. Finally, the design and synthesis of new commercial extractants (LIX, KELEX, SME, etc) specially suited for extraction and metal separation in hydrometallurgical processes has firmly contributed to a wide range of applications of metal extraction reactions [8].

The development of a practical application of any of these reagents necessitated a study of the behaviour of the system (feed solution and extractant + diluent), and of the relevant literature, leading to an enormous amount of diverse data (on distribution, spectroscopic, magnetic, colligative properties, etc.). Two main types of experimental information have been collected. Applied chemists and engineers were most interested in the quantitative aspects of the reactions, in the sense of describing the exctractive capacity of the reagents under extremely varied experimental conditions. Other investigators had more interest in the thermodynamic and physicochemical aspects of the reactions, especially for equilibrium and kinetics studies.

In this paper, the most usual experimental procedures, data collection, and graphical treatment in the study of metal extraction reactions are discussed. Curve-fitting treatment of metal distribution data is especially emphasized, with examples from our most recent work [11–14].

5.2   EXPERIMENTAL METHODS

Experiments on metal extraction systems are mainly designed to achieve the following objectives.

(i) The determination of the distribution coefficient of a given species under varied experimental conditions in order to determine the capacity of particular compositions of extractant + organic diluent for extracting metal ions.

(ii) The determination of the distribution coefficient of a given species under specific experimental conditions, in order to ascertain the compositions of the species extracted into the organic phase, and their formation constants.

The experimental determination of the distribution coefficient as a function of a master variable (pH, extractant concentration etc) in both type of work is an easy task which just requires the use of a well-tested analytical method (AAS, radio-tracers, spectrophotometry, etc) for the accurate determination of metal content in both phases, after equilibrium is attained.

From this information, the distribution constant for a given species M, defined by,

$$D_M = [M]_{tot,org}/[M]_{tot,aq} \qquad (5.1)$$

may be calculated.

Two precautions should be taken, however, in the determination of experimental distribution coefficients,

(i) Determination of the time for equilibrium.

Before the final data can be collected, the experimental function $D = f(t)$ must be constructed (Fig. 5.1). From this plot, the equilibrium position may be defined as the time for which the distribution coefficient remains constant within defined limits $\pm \delta D$.

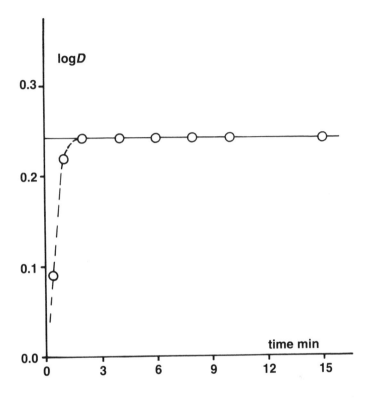

Fig. 5.1 — Log D plotted as a function of time. Data for the extraction of Co(II) in SCN$^-$ medium by TLAHCl in benzene [13].

(ii) Checking the mass balance conditions.

At equilibrium, the total number of moles of the distributed species must be equal to the sum of the number of moles in each phase. The following equation may be written:

$$n^*_{M_{tot,aq}} = n_{Mtot,aq} + n_{M_{tot,org}} \tag{5.2}$$

where n represents the number of moles and $n^*_{M_{tot,aq}}$ the total number of moles in the aqueous phase before the phases were in contact.

Taking into account phase volumes, Eq. (5.2) becomes

$$[M^*]_{tot,aq}V_{aq} = [M]_{tot,aq}V_{aq} + [M]_{tot,org}V_{org} \tag{5.3}$$

or, alternatively, as

$$[M^*]_{tot,aq} = [M]_{tot,aq} + V[M]_{tot,org} \tag{5.4}$$

where V represents the volumes ratio V_{org}/V_{aq}.

$[M^*]_{tot,aq}$ is the metal concentration in the feed solution and $[M]_{tot,aq}$ and $[M]_{tot,org}$ are known from analytical data, so the checking procedure is straightforward. In practical work, errors of around 2–3% are considered as satisfactory (Table 5.1).

Table 5.1 — Mass-balance equation for the extraction of Co(II) by the chelating reagent DMTADAP in MIBK at pH 9

Co(II)$_{tot,aq}$ (ppm)	Co(II)$_{tot,org}$	Co(II)$_{tot,aq}$	% Error
1.17	0.08	1.10	− 0.85
1.19	0.07	1.11	0.84
1.28	0.34	0.96	1.56
1.32	0.80	0.51	0.76
1.40	1.15	0.27	− 1.68
1.42	0.29	0.96	0.00

5.3 DATA TREATMENT IN PRACTICAL APPLICATIONS

This part refers mainly to distribution studies intended to obtain experimental information for practical applications. This type of work may include both analytical studies searching for optimal conditions for any metal determination, and laboratory studies necessary for designing a commercial process for metal extraction [8].

In this type of work, the distribution coefficient in Eq. (5.1) is frequently replaced by the function 'Percentage Extraction', defined by the following equation:

$$\%E = 100\ VD_m/(1 + VD_M) \tag{5.5}$$

which is plotted as a function of a suitable master variable (Fig. 5.2).

A similar function is utilized to express the capacity of a reagent for selective extraction of a cation mixture. In this case however, the separating factor, S defined by:

$$S = D_{M_I}/D_{M_{II}} \tag{5.6}$$

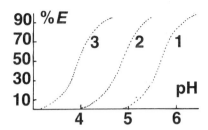

Fig. 5.2 — Percentage extraction of Cd(II) in nitrate medium by various extractants dissolved in toluene [12]. 1. 8-Methylphenylsulphonamidoquinoline = HR; 2. Mixture of HR and tributylphosphate (TBP); 3. Mixture of HR and tri-octylphosphine oxide (TOPO). C_S represents the concentration of the neutral organophosphorus compound.

gives quantitative information about the possibilities of metal separation. A high value of S indicates that $D_{M_I} \gg D_{M_{II}}$ and the species may be separated.

Working with mixtures of extractants can give rise to additional experimental information. In addition to distribution coefficients 'percentage extraction' and separating factors, a 'synergistic coefficient' may be determined (Figs. 5.2, 5.3). This

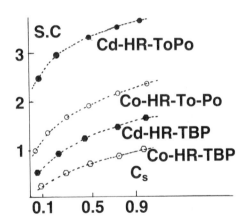

Fig. 5.3 — Experimental and theoretical synergic coefficient, Eq. (5.7) for the extraction of Cd(II) and Co(II) in nitrate medium by mixtures of extractants in toluene [12].

coefficient was defined by Taube and Siekierki [9] to be:

$$S.C. = \log [D_{1,2}/(D_1 + D_2)] \tag{5.7}$$

where D_1, D_2 and $D_{1,2}$ respectively represent experimental distribution coefficients when the organic phase contains either extractant E_1 or E_2 alone or a mixture. This gives quantitative information about the effect of the second extractant E_2 on the extraction of a given cation by the extractant E_1.

In analytical work, synergic coefficients have been estimated by the following equation [10]:

$$S.C = n \times \Delta pH_{50} \tag{5.8}$$

where n represents the charge on the metal ion, and ΔpH_{50} is the difference in pH corresponding to 50% extraction when the total concentration of the extractants both in the single systems and in the mixture are equal [11] (Table 5.2).

Table 5.2 — pH_{50} values for the extraction of Ni(II) with mixtures of the extractants DEHO and organophosphorus acids [11]

Extractant	pH_{50}	S.C.
DEHO–toluene	4.8	
DEHPA–toluene	5.1	
DOPA–toluene	6.1	
DEHO–DEHPA–toluene	0.2	9.2
DEHO–DOPA–toluene	1.1	7.3

$C_{DEHO} = C_{DEHPA} = C_{DOPA} = 0.05\ M$

5.4 STUDIES OF EQUILIBRIA

In the literature, the extraction of metal ions under varied experimental conditions has been explained in many cases by assuming that a single species is formed in the organic phase. Thus in the extraction of metal ions by acidic chelating reagents, reactions of the type:

$$M^{N+} + NHX\ \text{org} = MX_N\ \text{org} + NH^+ \tag{5.9}$$

have frequently been postulated.

The composition of the species extracted is found after graphical analysis of the distribution data by the slope analysis method. In this approach, the experimental function $\log D = f(\log[X])$ will give a straight line of slope N according to Eq. (5.10)

$$\log D = \log K_{MX_N} + NpH + N \log [HX] \tag{5.10}$$

More detailed experimental study of this type of system, however, soon showed that the single-species model is just a special case. The experimental $\log D = f(\log[X])$ functions are seldom straight lines if a range of experimental conditions is studied [11,16] (Fig. 5.4).

In fact, many results [11–14] indicate that both the metal ion and the extractant may participate in side-reactions, which must be included in the mass-balance equations in order to obtain a reliable estimate of the free extractant concentration in the organic phase.

Typical side-reactions are:

(i)   Self-association of organic extractants in the diluent.
(ii) Acid–base reactions of the extractant in the aqueous phase.
(iii) Complex-formation reactions of the metal ion with anionic components of the ionic medium.

Other complications may occur in metal extraction reactions. Solvation and polymerization in the organic phase seem to be frequent [15–17] and often the composition of the extracted species is strongly dependent on the experimental conditions. Species in the organic phase which contain the anionic form of the ionic medium have been described [18,19], and formation of mixed-complexes has been proposed as an explanation for synergism when mixtures of extractants are used [11,12,20–22].

It seems that two different types of experimental information are needed in order to be able to explain metal distribution data in terms of reactions and equilibrium constants.

Studies on the extractant.
This work should include:

(a) Determination of the state of aggregation of the extractant in the organic diluent,
(b) Determination of the distribution and acidity constants of the extractant,
(c) Identification of any chemical interactions in the organic phase in mixtures of extractants.

Metal extraction reactions
These studies should include:

(a) Extraction of metal ions by organic extractants (amines, chelating extractants, carboxylic acids, etc).
(b) Extraction of metal ions by mixtures of extractants (chelating amines, chelating extractants, neutral organophosphorous compounds, etc).

It is necessary to collect experimental data for a wide range of conditions. Good analytical work is necessary to obtain accurate distribution measurements. A wide range of concentrations of all components should be used in order to facilitate identification of all the species in solution.

We wanted to improve the standard experimental methods described in the literature. Therefore, for each of the systems studied, the following experimental details and procedures were followed.

(i) All reagents and solutions were purified and standardized.

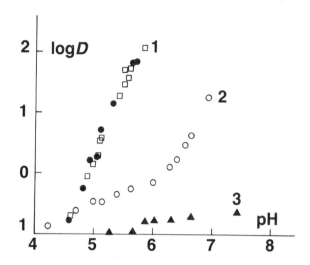

Fig. 5.4 — Examples of the experimental function $\log D = f(\text{pH})_{X_j}$: (1) Ni(II)–thiodecylacetic acid in toluene; (2) Ni(II)–5,8-diethyl-7-hydroxydodecane-6-oxime in toluene; (3) Ni(II)–n-hexanoic acid in toluene [11].

(ii) A medium of fixed ionic strength was used to ensure constant activity coefficients in the aqueous phase.
(iii) The concentration of extractant in the organic diluent was maintained below the limit for ideality.
(iv) Preliminary experimental measurements were made under diverse experimental conditions, in order to study the behaviour of the systems, and identify the region where equilibrium data would be taken.

5.5 DATA COLLECTION

Experimental data in liquid–liquid equilibria are generally of the type $D = f(X_i)_{X_j}$ in which the distribution coefficient of the metal is determined as a function of a given master variable. Nevertheless, the type of data to be collected depends fundamentally on the response of the system, and on the chemical characteristics of the extractant (that is, on the type of reaction to be studied).

In amine extraction, for instance, the following general reaction has been postulated [23–25].

$$M^{n+} + n\,X^- + r\,R_3NHX_{org} = MX_n(R_3NHX)_{r,\,org} \qquad (5.11)$$

where

$$K_{1nr} = [MX_n(R_3NHX)_r]_{org}/[M^{n+}][X^-]^n[R_3NHX]_{org}^r \qquad (5.12)$$

is the stoichiometric extraction constant.

The distribution coefficient of the species M is (cf. Eq. 5.1)

$$D = \sum_r K'_{1nr} [R_3NHX]^r_{org} \qquad (5.13)$$

where

$$K'_{1nr} = K_{1nr} [X^-]^n/(1+\Sigma\beta_i[X^-]^i) \qquad (5.14)$$

and β_i is the formation constant of the species MX_i in the aqueous phase. Equation (5.13) indicates that if no polynuclear species are present in the organic phase and the term $[X^-]$ is kept constant in the experiment, the distribution coefficient of the metal ion is a function only of the free amine salt concentration in the organic phase. For this case, data of the form $D = f([R_3NHX]_{org})_{[X^-]}$ should be collected. Then, the number of components in reaction (5.1) is effectively reduced and the determination of the different r values from mathematical analysis of Eq. (5.13) is made easier.

In the extraction of Ni^{2+} with the chelating extractant 7,8-diethyl-7-hydroxy-dodecan-6-one oxime (HL) [11], the following general reaction was postulated:

$$jNi^{2+} + (m+n) HL_{org} + rNO_3^- = Ni_jL_m(NO_3)_r(HL)_{n,\,org} + mH^+ \quad (5.15)$$

and the metal distribution coefficient (Eq. 5.1) was expressed as:

$$D = \sum_{jmrn} j\, K_{jmrn}[Ni^{2+}]^{(j-1)}[HL]^{(m+n)}_{org}\, [NO_3^-]^r\, h^{-m} \qquad (5.16)$$

This equation indicates that D depends on pH, free metal and nitrate ion concentrations, and on the free HL concentration in the organic phase.

Determination of the various stoichiometric coefficients and the values of the extraction constants K_{jmrn}, from Eq. (5.16) would require a set of experimental data of the type $D = f(X_i)_{X_j}$. Hence, using data of the type $f[Ni^{2+}]_{[NO_3^-], pH, [HL]_{org}}$, Eq. (5.16) may be written as:

$$D = \sum_j K_{jmrn} [Ni^{2+}]^{(j-1)} \qquad (5.17)$$

and the various j-values may be found. Mononuclear species, $j = 1$, should give D values that are independent of metal concentration in the aqueous phase.

On the assumption that only mononuclear species exist in the organic phase, ($j = 1$ in the previous treatment), Eq. (5.16) becomes:

$$D = \sum K_{lmrn} \, [\text{NO}_3^-]^r \, [\text{HL}]_{\text{org}}^{(m+n)} \, h^{-m} \tag{5.18}$$

and with a set of data of the type $D = f(\text{pH})_{[\text{NO}_3^-],[\text{HL}]_{\text{org}}}$, Eq. (5.18) can be transformed into function:

$$D = \sum_{lmrn} K'_{lmrn} \, h^{-m} \tag{5.19}$$

from which the stoichiometric coefficient m in Eq. (5.15) may easily be determined. Finally, from data for $D = f([\text{HA}]_{\text{org}})_{[\text{NO}_3^-],\,\text{pH}}$, Eq. (5.16) can be written as:

$$D = \sum_{lmrn} K''_{lmrn} \, [\text{HL}]^{(m+n)} \tag{5.20}$$

and the term $(m+n)$ may be determined from mathematical analysis of this function. In this type of treatment, however, the equilibrium concentration of HA in the organic phase cannot be obtained from any experimental measurement, and a function of the type $D = f([\text{HA}]_{\text{org}})$ at constant pH and NO_3^- concentration has to be determined graphically from the experimental function $D = f(\text{pH})_{[\text{NO}_3^-][\text{HL}]_{\text{org}}}$ by taking a section corresponding to constant pH.

Designing experimental methods from the general expression of the equation reactions often does not lead to the ideal experimental functions but, nevertheless may help in the acquisition of systematic information to clarify complicated equilibria.

5.6   TREATMENT OF DATA: GRAPHICAL METHODS

Most experimental data on distribution equilibria are treated by two different graphical methods. **Slope analysis** is used for preliminary treatment of $\log D = f([\text{X}])$ data, which is then, as the last step of the calculation, treated by **curve-fitting methods**.

In this graphical approach, introduced by Sillén and co-workers [26,27], experimental functions are compared with theoretical model functions representing different extraction reactions, until a fitting situation is found. Then, the mathematical form of the model function allows determination of the stoichiometric coefficients, and the equilibrium constants can be determined from the position of the function relative to the co-ordinate axes.

Although these graphical methods have been successful for interpretation of many types of data, they are unsuitable for analysing all the experimental information when more than three species are present. In this case, partial graphical analysis may be used to find the compositions and equilibrium constants of the predominant species.

The experimental function to be analysed varies with the different systems, but direct experimental data or an experimental function closely related to the measured variables usually offers, in our experience [11,12], the best chance of success.

The experimental and model functions are compared by translation along the co-ordinate axes. In the determination of a single stoichiometric coefficient and equilibrium constant, just a normalized variable is needed, and translation is made along one of the axes, keeping constant and identical positions in the other. When two equilibrium constants have to be calculated, translation must occur along both axes, and the shifts along the axes at the position of best fit will provide information for determination of the constants. Finally, if three species coexist in solution, the model function will depend on the value of a parameter L, and consequently it will consist of a family of curves. Here, model functions with different L values are compared, in the same manner as before, until a fitting position is found. At this position, both the L value and shifts along the axes will allow determination of the composition and formation constants of the various species extracted.

5.7 SELECTED EXAMPLES

In the following, details of the experimental procedures and graphical calculations for selected systems will be presented. Data on the extractant and metal–extractant systems are presented separately.

5.7.1 Studies on the extractant

Many organic extractants dissolved in diluents undergo self-association reactions of the type:

$$n \, B_{org} = B_{n \, org} \tag{5.21}$$

$$\beta_n = [B_n]/b^n \tag{5.22}$$

where B represents the extractant molecule (amine salts, long-chain carboxylic acids, alkyphosphoric acids, etc) and b is the free monomer concentration (Fig. 5.5).

Literature information on such reactions is scant and unreliable [7]. Nevertheless, recent studies by our group [11,15,28] have shown that careful experimental work can allow the determination of the degree of polymerization of many extractants.

5.7.1.1 *Self-association reactions in the system 7,8-diethyl-8-hydroxydodecan-6-one oxime in toluene [12,15]*

Experimental details. Solutions of the reagent B in toluene were prepared by weighing. Reactions were followed by vapour phase osmometry [16], and data of the type ($\Delta R, B$ in mole/kg) were collected.

The experimental variable ΔR, is related to the state of the solute in solution by the following equation:

$$\Delta R = K_1 \, S + K_2 \, S^2 \tag{5.23}$$

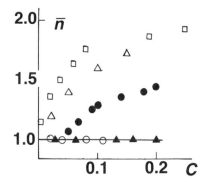

Fig. 5.5 — Mean aggregation numbers Eq. (5.25), for some organic solutions containing metal extractants (Osmometric data from [11,12,15,23]). □ Di-(ethylhexyl)phosphoric acid in toluene; △ Tri-n-octyl-ammonium hydrochloride in benzene; ● 7,8 diethyl-7-hydroxydode-can-6-one oxime; ▲ 8-methylphenylsulphonamidoquinoline in toluene. ○ benzil in toluene.

where K_1 and K_2 are constants which must be determined in separate experiments and S is the complexity sum, which represents the sum of the concentrations of all the species present. That is, according to (5.21) and (5.22),

$$S = b + \sum [B_n] = b + \sum_n \beta_n b^n \tag{5.24}$$

The constants in Eq. (5.23) are determined from data $\Delta R = f(C)$, where C represents a standard solution (benzil in toluene) which behaves as monomer over a wide concentration range. For this system, $S = C$ and K_1 and K_2 may be calculated by a least-squares treatment (Table 5.3). Alternative plots $\Delta R/C = f(C)$ (Fig. 5.6) allow

Table 5.3 — Calibration constants in Eq. (5.23) with benzil in toulene as standard

Method	K_1	K_2
Graphical	2142	− 906
Letagrop–SUMPA	2143 ± 5	− 913 ± 23

graphical estimation of both constants from the slope and intercept of the straight line.

Treatment of the data: $(\Delta R, B)$. From a set of data $(\Delta R, B)$, the mean aggregation number $\bar{n}$, defined by,

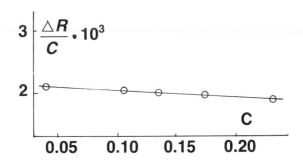

Fig. 5.6 — The function $\Delta R/C$ for benzil in toluene for graphical determination of constants in Eq. (5.23).

$$\bar{n} = B/S = \frac{\sum\limits_{n} n\,[B_n]}{\sum\limits_{n}[B_n]} = \frac{\sum\limits_{n} n\,\beta_n\,b^n}{\sum\limits_{n} \beta_n\,b^n} \tag{5.25}$$

was calculated (Table 5.4) and from this, the free monomer concentration was

Table 5.4 — Details of the calculation of free monomer concentration in self-association equilibria [12]

B, mole/kg	S	$\bar{n}$	b
0.008	0.008	1.00	0.008
0.019	0.018	1.02	0.018
0.032	0.031	1.05	0.030
0.084	0.068	1.23	0.061
0.116	0.089	1.31	0.075
0.182	0.129	1.41	0.098
0.258	0.171	1.51	0.116

computed by means of the following equation:

$$\log b = \log b_0 + \int_{S_o}^{S} (\bar{n})^{-1}\, d(\log S) \tag{5.26}$$

(Table 5.4, c.f. Section 4.2, p. 50).

Determination of the composition of the aggregates. In order to determine the composition of the various aggregates, i.e. the various n-values in Eq. (5.21), and the values of the self-association constants, let us write the mass-balance equation for solute B:

$$B = b + \sum_n n\,[B_n] = b + \sum_n n.\beta_n\, b^n \tag{5.27}$$

and let us assume the formation of a unique aggregate of composition B_N. In this case, Eq. (5.27) becomes

$$\log (B/b - 1) = N \log \beta_N + (N-1)\log b \tag{5.28}$$

Since the terms B and b are known for each experimental point, values for log $(B/b - 1)$ may be calculated and plotted as a function of the free monomer concentration (Fig. 5.7). The slope of this straight line indicates that $N - 1 = 2$, so it seems possible that the species B_3 may be present. The value of the trimerization constant β_3 may be estimated from the intercept in Fig. 5.7.

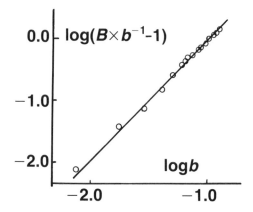

Fig. 5.7 — The experimental function log $(Bb^{-1} - 1)$ as a function of free monomer concentration. Continuous line represents a straight line of slope $(N-1) = 2$, Eq. (5.28).

Alternative graphical method. The results of the preliminary treatment suggest that the experimental data may be explained by assuming the existence of a monomer–n-mer equilibrium; therefore let us make a graphical search for the most probable n-mer.

On this assumption, Eq. (5.27) becomes:

$$B/b = 1 + N \beta_N b^{N-1} \tag{5.29}$$

which indicates that plots of the type $B/b = f(b)^{(N-1)}$ will be straight lines of slope N and intercept equal to 1, if a single polymeric species exists at equilibrium. This is shown in Fig. 5.8a,b,c; the results support the results of the previous treatment, which postulated the existence of a trimeric molecule.

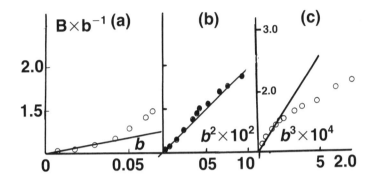

Fig. 5.8 — The experimental function $Bb^{-1} = f(b)$. The straight lines represent the following reaction models: (a) monomer–dimer; (b) monomer–trimer; (c) monomer–tetramer.

Curve-fitting treatment. The plot in Figs. 5.8b shows deviation around $b = 0.05$ which suggested the possible existence of a dimer in this concentration zone. Because of this, a model assuming the existence of a monomer–dimer–trimer equilibrium was tested [12].

For this reaction model, Eq. (5.27) may be written as:

$$(B - b)/(2\beta_2 b^2) = 1 + \tfrac{3}{2} \beta_3 \beta_2^{-1} b \tag{5.30}$$

and defining new variables as:

$$Y = (B - b)/(2 \beta_2 b^2) \tag{5.31}$$

$$X = \tfrac{3}{2} \beta_3 \beta_2^{-1} b \tag{5.32}$$

the model function

$$\log Y = \log (1 + X) \tag{5.33}$$

was obtained.

This model function was compared with the experimental function $\log(B - b)/b^2 = f(\log b)$ in Fig. 5.9. The goodness of the fit confirms the proposed

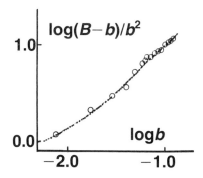

Fig. 5.9 — Experimental function $\log(B - b)/b^2 = f(b)$ compared with the model function in Eq. (5.33) in the 'best fitting position'.

monomer–dimer–trimer equilibrium. Finally, according to Eqs. (5.31) and (5.32), the shifts along the axes at the best fitting position

$$\log Y - \log (B - b)/b^2 = -\log 2\, \beta_2 \tag{5.34}$$

$$\log X - \log b = \log 3\, \beta_3 - \log 2\, \beta_2 \tag{5.35}$$

provides enough information for the determination of the values of the association constants β_2 and β_3.

5.7.1.2 *Distribution and acid–base equilibria of the molecular form of a chelating extractant*

This type of equilibrium, which in the simplest of the cases may be represented by

$$\text{HA} \rightleftharpoons \text{HA}_{\text{org}}; \quad K_d = [\text{HA}]_{\text{org}}/[\text{HA}] \tag{5.36}$$

$$\text{HA} \rightleftharpoons \text{A}^- + \text{H}^+; \quad K_a = [\text{A}^-]\, h/[\text{HA}] \tag{5.37}$$

has been studied for soluble chelating agents such as azo-derivatives [30,31]. Here, the determination of the constants K_d and K_a is a relatively easy task, that requires accurate determination of the pH and total extractant concentrations in the two phases, at equilibrium. Determination of the distribution coefficient is facilitated in those cases when the chelating reagent absorbs light.

The problem may be more difficult for commercial extractants and synthetic model reagents [11,12], first, because the low solubility of a reagent of moderate molecular weight results in solutions of low buffer capacity in which it is difficult to make accurate pH measurements. Secondly, the lack of any specific properties (electrical, optical, etc) may make it difficult to find a reproducible method for the determination of the distribution coefficients.

A model experiment: determination of the distribution and acidity constant of the synthetic extractant 8-methylphenylsulphonamidoquinoline [12]. This reagent, synthesised as a model reagent for studying the behaviour of commercial extractant LIX-34 [12], had a spectrum which clearly indicated the possibility of analytical determination of the species HA. Aqueous-alcohol solutions of the reagent obeyed Beer's law for $\lambda = 235$ nm (Fig. 5.10). pH measurements in acidic and basic conditions presented no difficulty, and a set of data of the form $D = f(\text{pH})_{[\text{HA}]_{\text{tot, org}}}$ was collected (Fig. 5.11)

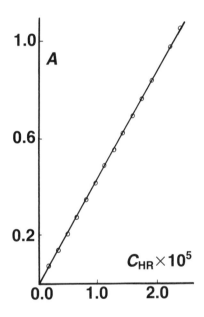

Fig. 5.10 — Absorbance of aqueous alcohol solutions of the reagent 8-methylphenylsulphona-midoquinoline (HR) at pH = 1, $\lambda = 235$ nm.

Simultaneous determination of the constants K_d, K_{a1} and K_{a2}. The form of the curves in Fig. 5.11 indicated the possible existence of the following equilibria;

$$HA \rightleftharpoons HA_{org} \qquad ; K_d$$

$$H_2A^+ \rightleftharpoons HA + H^+ \qquad ; K_{a1}$$

$$HA \rightleftharpoons A^- + H^+ \qquad ; K_{a2}$$

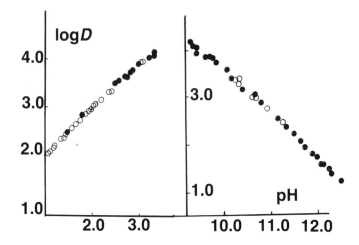

Fig. 5.11 — Distribution coefficient of the reagent HR as a function of pH.

The distribution coefficient of the species HA may then be written as:

$$D = \frac{[HA]_{tot,org}}{[HA]_{tot,aq}} = K_d/(1 + K_{a1}^{-1} h + K_{a2} h^{-1}) \tag{5.38}$$

Eq. (5.38) may be transformed as follows:

$$D^{-1} K_d K_{a2}^{-1} h = 1 + K_{a2}^{-1} h + K_{a1}^{-1} K_{a2}^{-1} h^2 \tag{5.39}$$

As three equilibrium constants must be simultaneously determined, defining the new variables:

$$Y = D^{-1} K_d \cdot K_{a2}^{-1} h \tag{5.40}$$
$$X = K_{a2}^{-1} h \tag{5.41}$$

and the parameter

$$L = K_{a1}^{-1} K_{a2} \tag{5.42}$$

leads to the following model function:

$$Y = 1 + X + L X^2 \tag{5.43}$$

This equation indicates that the model function will have different forms for different

L values. Hence, from comparisons of the function $\log Y = f(\log X)$ for different L values, with the experimental function $\log Dh^{-1} = f(\text{pH})$, a fitting position was found in Fig. 5.12. In this position the shifts along the axes give values that can be

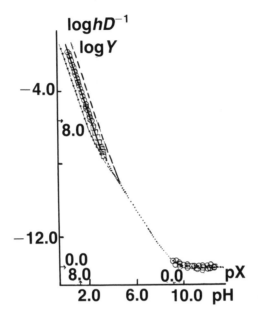

Fig. 5. 12 — Experimental function $\log D^{-1}h = f(\text{pH})$ compared with the model function in Eq. (5.43) for various values of L. ●●●●● $L = 1.6 \times 10^{-7}$; —— $L = 1.6 \times 10^{-6}$; – – – $L = 1.6 \times 10^{-5}$.

substituted into the following, logarithmic forms of Eqs. (5.40) and (5.41).

$$\log Y + \log Dh^{-1} = \log K_d + pK_{a2} \tag{5.44}$$
$$pX - pH = -pK_{a2} \tag{5.45}$$

From the best value for the parameter L, (Fig. 5.12)

$$\log L = pK_{a1} - pK_{a2} \tag{5.46}$$

Solving Eqs. (5.44–5.46) yields values for the constants K_d, K_{a1} and K_{a2}.

5.7.2 Metal-extraction reactions

Metal extraction reactions may be represented by the following general equation:

$$jM^{n+}(m+n)HA_{org} + r\,X^- \rightleftharpoons M_jA_m(HA)_nX_{r,\,org} + mH^+ \tag{5.47}$$

where X^- is the anionic form of the ionic medium and HA is a chelating extractant. When a moderate concentration of this reagent is used together with an ionic medium of constant strength, the activities of the various species are practically constant and the stoichiometric extraction constant, K_{jmnr} may be defined as:

$$K_{jmnr} = [M_jA_m(HA)_mX_r]_{org} \, h^m / ([M^{n+}]^j \, [HL]_{org}^{(m+n)} \, [X^-]^r) \qquad (5.48)$$

The object of the graphical analysis is to determine the values of the stoichiometric coefficients j, m, n and r, and the stoichiometric equilibrium constants. Because metal extraction reactions may vary with working conditions (nature of the extractants and metal ion, composition of phases in contact, etc) there is no standard graphical procedure for data treatment, and graphical analysis must be done as a step-wise procedure. In the following, some examples of different types of model calculations will be presented.

5.7.2.1 Studies on the extraction of copper(II) with 2-(ethylhexyl)phosphoric acid (DEHPA) in the commercial diluent Isopar-H [32]

Experimental details. The phases in contact have the following compositions:

Aqueous phase: $(Na^+, H^+, Cu^{2+}) ClO_4^-$,   1.0 M

$(Na^+, H^+, Cu^{2+}) SO_4^{2-}$,   1.0 M

Organic phase: DEHPA in Isopar- H

0.01 ⩽ [DEHPA] ⩽ 0.2 M

At equilibrium, the Cu(II) concentration in both phases was determined by AAE, and the distribution coefficient of the metal, defined by:

$$D = [Cu]_{tot,org}/[Cu]_{tot,aq} \qquad (5.49)$$

was calculated. As the compositions of the phases were varied, data of the form $D = f(pH)_{HA}$ and $D = f(DEHPA)_{Cu,pH}$ were collected.

Treatment of the data. The extraction of Cu(II) by 2-(ethylhexyl)phosphoric acid has been represented as follows:

$$Cu^{2+} + mHA_{org} = CuA_2(HA)_{m-2,\,org} + 2H^+ \qquad (5.50)$$

with stoichiometric extraction constant given by

$$K_{lm} = [CuA_2(HA)_{m-2}]_{org} \, h^2 / [Cu^{2+}][HA]_{org}^m \qquad (5.51)$$

The distribution coefficient for Cu(II) is therefore

$$D = [CuA_2(HA)_{m-2}]_{org}/[Cu(II)]_{aq} = K_{lm}[HA]_{org}^m\, h^2/\gamma \qquad (5.52)$$

where

$$\gamma = 1 \text{ for ClO}_4^- \text{ medium}$$

$$\gamma = 1 + (\beta_1[SO_4]_{tot}/(1 + h/K_{a2})) \text{ for sulphate medium}$$

β_1 is the formation constant of $CuSO_4^+$ and K_{a2}, the dissociation constant of HSO_4^-.
  In order to determine the stoichiometric coefficient m, the number of DEHPA molecules solvating the neutral complex CuA_2, Eq. (5.52) is written as follows:

$$\log D - 2\text{ pH } - \log \gamma = \log K_{lm} + m \log[HA]_{org} \qquad (5.53)$$

and taking into account the dimerization equilibrium of the extractant,

$$2HA_{org} = (HA)_{2,\,org}\,, \qquad K_2 = [(HA)_2]_{org}/[HA]_{org}^2 \qquad (5.54)$$

Eq. (5.53) becomes:

$$\log D - 2\text{pH } - \log \gamma = \log K_{lm} - (m/2)\log K_2 + (m/2)\log[(HA)_2]_{org} \qquad (5.55)$$

As the existence of a single species of composition $CuA_2(HA)_{m-2}$ has been assumed, plots of the function $\log D - 2$ pH $+ \log \gamma$ as a function of the dimer concentration should provide straight lines of slope $m/2$. These plots in Fig. 5.13

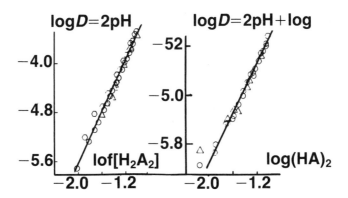

Fig. 5.13 — The function $\log D - 2$ pH $+ \log \gamma$ plotted as a function of the concentration of dimer in the organic phase. The coincidence of points for different metal concentrations ([Cu(II)]$= \bigcirc$, 0.01, $\triangle$ 0.02 and $\bigcirc$ 0.03 mole/l) shows that the species extracted is mononuclear.

indicate that the experimental data fit acceptable straight lines of slope 2. According to Eqs. (5.55) and (5.50), the results indicate that four HA groups are bound to Cu(II) in the extracted species. From previous data it was demonstrated that 2 protons were displaced in Eq. (5.50), so the complex extracted may have the composition $CuA_2(HA)_2$. The equality of slopes in Fig. 5.13 indicates that neither ClO_4^- nor SO_4^{2-} participate in the composition of the species in the organic phase, and the only medium effect on complexation in the aqueous phase was easily corrected by means of the term $\log \gamma$, (Eq. (5.52)).

In order to calculate the concentration of the dimer in the organic phase, the following equation was used:

$$[HA]_{tot,org} = [HA]_{org} + 2[(HA)_2]_{org} + m[CuA_2(HA)_2]_{org} + [HA] + [A^-] \quad (5.56)$$

and, from Eq. (5.54),

$$[HA]_{tot,org} = K_2^{-1/2}[(HA)_2]_{org}^{1/2} + 2[(HA)_2]_{org} + [Cu_2A_2(HA)_2]_{org} + [HA] + [A^-] \quad (5.57)$$

Because of the low solubility of DEHPA in the aqueous phase and the high value of the dimerization constant, $K_2 = 10^{4.7}$ Eq. (5.57) may be written as:

$$[(HA)_2]_{org} = \frac{[HA]_{tot,org}}{2} - \frac{mD}{2(1+D)} [Cu(II)]_{tot} \quad (5.58)$$

where $[Cu(II)]_{tot}$ is the total copper concentration in the two phases when the phase ratio (V) is unity. Because Cu(II) is present in macro quantities, it cannot be neglected in Eq. (5.58) and the dimer concentration has to be calculated for each different value of m.

5.7.2.2 *The system nickel(II)–2-(ethylhexyl)phosphoric acid in toluene* [11,18]
The extraction of Ni(II) by DEHPA may be represented by the following reaction:

$$Ni^{2+} + (p+q)/2 \ (HA)_{2,\,org} = NiA_p \ (HA)_{q,\,org}^{2-p} + p \ H^+ \quad (5.59)$$

in which both the values for the stoichiometric coefficients p, q and the extraction constant K_{pq} must be determined.

Experimental details: the phases in contact had the following compositions:

Aqueous phase:: (Na^+, H^+, Ni^{2+}) NO_3^-, 1.0 M

Organic phase: : DEHPA in toluene

$$0.01 \leqslant [DEHPA] \leqslant 0.200 \ M$$

After equilibrium was attained, Ni(II) was determined in both phases by AAE, and pH values were measured. By changing the H^+ concentration in the test solution, a set of data of the type $D_{Ni} = f(pH)_{HA}$ was collected (Fig. 5.14).

Curve-fitting analysis. The stoichiometric extraction constant may be written as:

$$K_{pq} = [NiA_p(HA)_q]_{org} \, h^p / [Ni^{2+}][(HA)_2]_{org}^{(p+q)/2} \tag{5.60}$$

and, the distribution coefficient of the metal may be expressed as:

$$D = \sum_p \sum_q K_{pq} \, \alpha_{Ni}^{-1} \, [(HA)_2]_{org}^{(p+q)/2} \, h^{-p} \tag{5.61}$$

where α_{Ni} is the side-reaction coefficient of Ni(II) in the aqueous phase, given here by:

$$\alpha_{Ni} = 1 + \beta_1 \, [NO_3^-] + \beta_2 [NO_3^-]^2 \tag{5.62}$$

where the β_n are the formation constants of the Ni(II)–NO_3^- complexes.

Determination of number of protons lost, p. In order to determine the coefficient p in Eq. (5.59), let us assume just a single species of composition $NiA_P(HA)_Q$ is formed. In this case Eq. (5.61) becomes:

$$D = K_{PQ} \, \alpha_{Ni}^{-1} \, [(HA)_2]_{org}^{(P+Q)/2} \, h^{-P} \tag{5.63}$$

which may be transformed into the model function:

$$D = X^{-P} \tag{5.64}$$

in which

$$X = (K_{PQ} \, \alpha_{Ni}^{-1} \, [(HA)_2]_{org}^{(P+Q)/2})^{-1/P} \, h \tag{5.65}$$

By comparing data in Fig. 5.14 with the model function $\log D = f(pX)$ for different P values, by translation along the abscissa, a fitting position was found for $P = 2$. This indicates that 2 protons are released in Eq. (5.59), so the species extracted may have the composition $NiA_2(HA)_Q$ (Fig. 5.14).

Determination of Q and K_{PQ}. According to Eq. (5.65) the difference in the abscissa in the position of best fit, which depends on dimer concentration, may be expressed as follows:

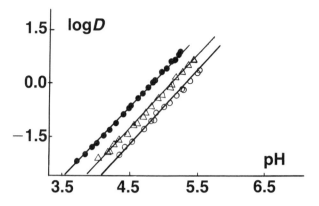

Fig. 5.14 — Experimental values for $\log D = f(\text{pH})_{\text{DEHPA}}$ at different constant concentrations of the extractant. Straight lines represent the model function in Eq. (5.64). $C_{\text{DEHPAorg}} = \bullet, 0.096; \triangle, 0.048$ and $\bigcirc, 0.027\ M$.

$$\log X + \text{pH} = -1/P\{\log(K_{PQ}\ \alpha_{\text{Ni}}^{-1}) + [(P+Q)/2]\ \log\ [(HA)_2]_{\text{org}}\} \quad (5.66)$$

Hence, from differences in Fig. 5.14, at the various $[HA]_{\text{tot}}$ values, a set of data of the type $\log X + \text{pH} = f([(HA)_2]_{\text{org}})$ was obtained and plotted in Fig. 5.15.

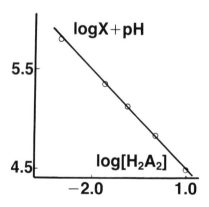

Fig. 5.15 — Differences in abscissa in Eq. (5.66) plotted as a function of the dimer concentration in the organic phase.

The function $\log X + \text{pH}$ yields a straight line of slope -1, which indicates according to Eq. (5.66)

$$(P+Q)/2P = 1 \quad \text{and} \quad Q = 2 \tag{5.67}$$

From this result, the existence in the organic phase of a species of composition $NiA_2(HA)_2$ may be postulated.

Finally, the formation constant of this species, [Eq. (5.60)] was determined from the intercept of the straight line [Fig. 5.15, Eq. (5.66)]. The dimer concentration in the organic phase was estimated from the distribution, dimerization and acid–base constants for the reagent in the system 1.0 M (Na^+, H)ClO_4^-/DEHPA in toluene [12].

5.7.2.3 Polymerization of metallic species in the organic phase: studies on the extraction of nickel(II) by n-dodecanoic acid in toluene [33]

Experimental details. The phases in contact had the following compositions:

$$\text{Aqueous phase}: (Na^+, H^+, Ni^{2+})NO_3,\ 2.0\ M$$

$$\text{Organic phase}\ :\ \text{n-dodecanoic acid in toluene}$$

$$0.05 \leqslant [HA] \leqslant 0.3\ M$$

After equilibrium had been attained, [Ni(II)] and pH were determined in both phases. By keeping constant, at different levels, the total concentration of the carboxylic acid, and varying the pH of the aqueous phase, a set of data of the type $D_{Ni} = f(pH)_{HA}$ was collected.

Treatment of the data: the extraction of Ni^{2+} by the reagent HA may be described by the following general reaction:

$$j\ Ni^{2+} + \tfrac{1}{2}(2+m)j\ (HA)_{2,\,org} = (NiA_2(HA)_m)_{j,\,org} + 2\,j\ H^+ \qquad (5.68)$$

where HA represents n-dodecanoic acid, which behaves as dimer in the organic diluent, and K_{mj} the stoichiometric extraction constant, defined by:

$$K_{mj} = [(NiA_2(HA)_m)_j]_{org}\ h^{2j}/[Ni^{2+}]^j\ [(HA)_2]_{org}^{\frac{1}{2}(2+m)j} \qquad (5.69)$$

Then, from Eqs. (5.62) and (5.69), the distribution coefficient for Ni(II) is:

$$D = \sum_j \sum_m jK_{mj}\ \alpha_{Ni}^{-1}\ [Ni^{2+}]^{(j-1)}\ [(HA)_2]_{org}^{\frac{1}{2}(2+m)j} \times h^{-2j} \qquad (5.70)$$

The two last equations will be utilized in the determination of the various stoichiometric coefficients m, j and constants K_{mj}.

Preliminary determination of j. In order to estimate the possible j values, the mass-balance equation for Ni(II) in the organic phase may be written as:

$$C_{\text{Ni,org}} = \sum_j \sum_m j \, [(\text{NiA}_2(\text{HA})_m)_j]_{\text{org}} \tag{5.71}$$

and taking into account Eq. (5.69),

$$C_{\text{Ni,org}} = \sum_j \sum_m j \, K_{mj} \, [\text{Ni}^{2+}]^j \, [(\text{HA})_2]_{\text{org}}^{\frac{1}{2}(2+m)j} \, h^{-2j} \tag{5.72}$$

Now, if the existence in the organic phase of a single species of composition $(\text{NiA}_2(\text{HA})_M)_J$ is assumed, Eq. (5.72) becomes:

$$\log C_{\text{Ni,org}} = \log J \, K_{MJ} + \tfrac{1}{2}(2+M) \, J \, \log[(\text{HA})_2]_{\text{org}} + J(\log [\text{Ni}^{2+}] + 2 \, \text{pH}) \tag{5.73}$$

This equation indicates that plots of the experimental functions $\log C_{\text{Ni,org}} = f(\log [\text{Ni}^{2+}] + 2 \, \text{pH})$ will be straight lines if the previous hypothesis is correct.

The results of this analysis are shown in Fig. 5.16, which indicates that straight

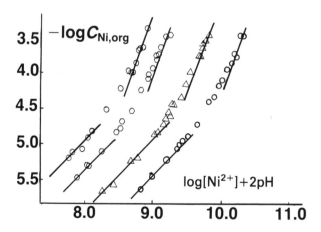

Fig. 5.16 — Total Ni(II) concentration in the organic phase plotted as a function of $\log[\text{Ni}^{2+}] + 2 \, \text{pH}$. Straight lines of slope 1 and 3 are fitted to the experimental data.

lines of different slopes fit the data in the different concentration regions. Hence, it must be concluded that several species, of composition $(NiA_2(HA)_m)_j$ with $1 \leqslant j \leqslant 4$ must exist.

Curve fitting analysis. Taking into account the results of this previous analysis, the existence of two species of general composition $(NiA_2(HA)_M)$ and $(NiA_2(HA)_N)_J$ may be assumed. On this assumption, Eq. (5.70) becomes:

$$D = K_{M1} \alpha_{Ni}^{-1} [(HA)_2]_{org}^{\frac{1}{2}(M+2)} h^{-2}$$
$$+ JK_{NJ} [Ni^{2+}]^{(J-1)} [(HA)_2]_{org}^{\frac{1}{2}(N+2)J} h^{-2J} \alpha_{Ni}^{-1} \qquad (5.74)$$

where K_{M1} and K_{NJ} are the corresponding extraction constants defined by Eq. (5.69). This equation may be written as follows:

$$D h^2 \alpha_{Ni}/K_{M1} [(HA)_2]_{org}^{(M+2)/2}$$
$$= 1 + JK_{NJ} [Ni^{2+}]^{(J-1)} [(HA)_2]_{org}^{-1-M/2+[J(1+N/2)]}/[K_{M1} h^{-2(J-1)}] \quad (5.75)$$

To determine the J-value, the following model function is defined:

$$Y = \log (1 + X^i) = f(\log X) , \qquad i = J - 1 \qquad (5.76)$$

where

$$Y = \log D - 2 \text{ pH} + \log \alpha_{Ni} - \log K_{M1} - (M+2)/2 \log[(HA)_2]_{org} \quad (5.77)$$
$$\log X = (\log [Ni^{2+}] + 2 \text{ pH}) + (J-1)^{-1} \log J K_{NJ} K_{M1}^{-1}$$
$$+ (J-1)^{-1} (-1 - M/2 + J(1-N/2)) \log [(HA)_2]_{org} \qquad (5.78)$$

which was compared for various i values with the experimental functions $\log D - 2 \text{ pH} + \log \alpha_{Ni} = f(\log[Ni^{2+}] + 2 \text{ pH})$ in Fig. 5.17a–c.

From these plots, it is evident that the function $Y = \log (1 + X^2)$ represents the best fit to the data. From the value of $i = J - 1 = 2$, it may be concluded that species of composition $NiA_2(HA)_M$ and $(NiA_2(HA)_N)_3$ are predominant in the organic phase.

Determination of M and N and the extraction constants K_{M1} and K_{N3}. Equations (5.77) and (5.78) indicate that the differences in the co-ordinate axes may be expressed as a function of the dimer concentration,

$$\Delta Y = (\log D - 2 \text{ pH} + \log \alpha_{Ni}) - Y =$$
$$= \log K_{M1} + (M+2)/2 \log [(HA)_2]_{org} \qquad (5.79)$$
$$\Delta X = \log X - (\log[Ni^{2+}] - 2 \text{ pH}) = 1/2 \log 3K_{N3} K_{M1}^{-1}$$
$$+ \frac{1}{2}(-1 - M/2 + 3(1 - N/2)) \log [(HA)_2]_{org} \qquad (5.80)$$

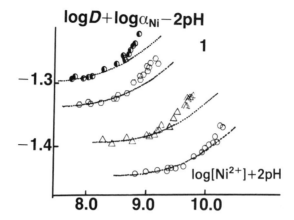

Fig. 5.17 — Experimental functions $\log D + \log \alpha_{Ni} - 2$ pH plotted against $\log[\text{Ni}^{2+}] + 2$ pH, compared with model function in Eq. (5.76) for various i values. The following models were tested: 1. monomer–dimer ($i = 1$); 2. monomer–trimer ($i = 2$); 3. monomer–tetramer ($i = 3$).

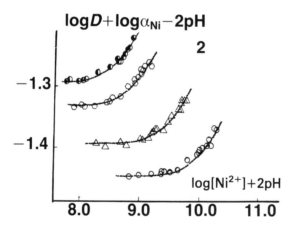

Fig. 5.17.

Hence, in order to determine the coefficients M and N, values of ΔY and ΔX were calculated from the plots in Fig. 5.17, (Table 5.5) and plotted as a function of the dimer concentration. The results (Fig. 5.18) show good linearity of both functions. The slopes of the two lines allow the determination of the coefficients $M = 2$ and $N = 2$, and the extracted species may have the compositions $\text{NiA}_2(\text{HA})_2$ and $(\text{NiA}_2(\text{HA})_2)_3$. From the intercepts of the lines, the values of the extraction constants K_{21} and K_{23} were determined (Table 5.7).

Finally, the dimer concentration was determined from the following equation, taking the phase ratio $V = 1$:

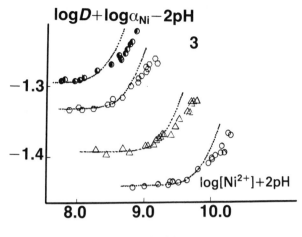

Fig. 5.17.

Table 5.5 — Differences in co-ordinates ΔY and ΔX, according respectively to Eqs. (5.79) and (5.80) as a function of the concentration of dimer in the organic phase [11].

ΔY	ΔX	$\log[(HA)_2]org$
-14.4	-9.9	-1.60
-13.9	-9.5	-1.30
-13.3	-8.8	-1.00
-12.9	-8.6	-0.82

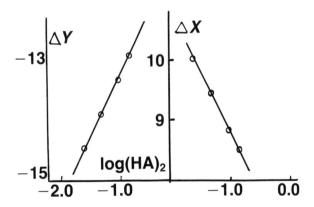

Fig. 5.18 — Differences in co-ordinates obtained from Fig. 5.17 (2) in the best fitting position, plotted as a function of the dimer concentration in the organic phase.

Table 5.6 — Values for association constants for the system 5,8-diethyl-7-hydroxydodecan-6-one oxime in toluene

Method	$\log \beta_2$	$\log \beta_3$
Graphical	0.38	1.38
Letagrop–SUMPA	0.45 ± 0.19	1.31 ± 0.05

Table 5.7 — Composition of the species and their formation constants. System: Ni(II)–n-dodecanoic acid–toluene

Graphical	$\log K_{mj}$	Letagrop–DISTR	$\log K_{mj}$
		NiA_2	-12.92 ± 0.12
$NiA_2(HA)_2$	-11.52	$NiA_2(HA)_2$	-11.52 ± 0.18
$(NiA_2(HA)_2)_2$	-19.10	$(NiA_2(HA)_2)_2$	-18.96 (max) to -18.67
$(NiA_2(HA)_2)_3$	-25.55	$(NiA_2(HA)_2)_3$	-26.57 (max) to -26.30

$$C_{HA,org} = [HA]_{org} + 2[(HA)_2]_{org} + C_{HA,aq}$$

$$+ \sum_j \sum_m (m+2) \, j \, [(NiA_2HA)_m)_j]_{org} \tag{5.80}$$

Since n-dodecanoic acid was in excess in the organic phase, the last term of this equation may be neglected. Also, the low solubility of the carboxylic acid in the aqueous phase, and its great tendency to dimerize in the organic diluent, might suggest that it is possible to neglect the terms $C_{HA_{aq}}$ and $[HA]_{org}$ in Eq. (5.81) which then would become:

$$C_{HA,org} = 2 \, [(HA)_2]_{org} \tag{5.82}$$

which makes the proposed calculations seem futile.

5.8   CONCLUDING REMARKS

The experimental methods and graphical calculations included in this paper represent selected pieces of an extensive work aiming toward the understanding of the chemical reactions which explain metal extraction processes [11–14,34].

The work, which included studies of long-chain amines and carboxylic acids, neutral and acidic organophosphorus compounds and chelating extractants, has shown that collection of accurate and extensive experimental information is a necessary step in studies of metal distribution equilibria. The different types of calculations indicate the great versatility of curve-fitting methods, and also the limitations of the slope analysis method Eq. (5.10), used by many authors.

Curve-fitting methods are useful as an independent tool for the determination of the equilibrium composition of the organic phase, expressed in terms of the stoichiometry of the species extracted and their formation constants. Comparisons in Tables 5.6–5.8 reinforce this statement, mainly because speciation models determined by the two methods are practically coincident.

Table 5.8 —Survey of the species and extraction constants for the system Ni(II)–5,8-diethyl-7-hydroxydo-decan-6-one oxime in toluene. $0.010 \leqslant C_{HL} \leqslant 0.1\ M$.

Graphical	$\log K_{rmn}$	Letagrop–DISTR	$\log K_{rmn}$
NiL$_2$	-9.47	NiL$_2$	-9.50 ± 0.15
Ni(NO$_3$)L(HL)$_2$	-1.21	Ni(NO$_3$)L(HL)$_2$	-0.92 ± 0.18
Ni(NO$_3$)$_2$(HL)	1.53	Ni(NO$_3$)$_2$(HL)	1.71 ± 0.10

Two main factors make curve-fitting so effective. First, the ability of model functions to adjust to the experimental information and make possible the selection of the more probable species (Figs. 5.8 and 5.17) and the fact that the quality of the data determines the form of the experimental functions and consequently, the goodness of fit and the accuracy of the equilibrium constants (Tables 6–8). The data in these tables show that the final refinement of the graphical models by the program LETAGROP-DISTR should be an easy task [34].

Acknowledgements
I wish to thank all the co-workers who have contributed to this work, especially Dr. Manuel Valiente of the Universitat Autónoma de Barcelona, Dr. Maria Elizalde and Dr. Jose Maria Castresana of the Universidad del Pais Vasco, Bilbao, and Dr. Ana Maria Sastre, Dr. Juan de Pablo, Dr. N. Miralles and I. Casas of the Universitat Politècnica de Cataluya (UPC), Barcelona.

REFERENCES

[1] C. D. Braun, *Z. Anal. Chem.*, 1863, **2**, 36.
[2] C. D. Braun, *Z. Anal. Chem.*, 1867, **6**, 86.
[3] W. Nernst, *Z. Phys. Chem. (Leipzig)*, 1891, **8**, 110.
[4] G H. Morrison and H. Freiser, *Solvent Extraction in Analytical Chemistry*, Wiley, New York, 1957.
[5] A. K. De, S. M. Khopkar and R. A. Chalmers, *Solvent Extraction of Metals*, Van Nostrand–Reinhold, London, 1970.
[6] T. Sekine and Y. Hasegawa, *Solvent Extraction Chemistry, Fundamentals and Applications*, Dekker, New York, 1977.
[7] Y. Marcus and A. S. Kertes, *Ion Exchange and Solvent Extraction of Metal Complexes*, Wiley, New York, 1969.
[8] D. S. Flett, J. Melling and M. Cox, in *Commercial Solvent Systems for Inorganic Processes. Handbook of Solvent Extraction*, M. H. Baird and C. Hanson eds., Wiley, 1983.
[9] M. Taube and S. Siekierki, *Nukleonica*, 1961, **6**, 489.
[10] G. Duyckaerts and J. F. Desreux, *ISEC 77, International Solvent Extraction Conference, Proceedings*, Toronto, Canada, 1977.
[11] M. P. Elizalde, *Doctoral Thesis*, Universidad del Pais Vasco, Servicio Editorial, Lejona, Spain, 1982.
[12] J. M. Castresana, *Doctoral Thesis*, Universidad del Pais Vasco, Bilbao, Spain, 1983.
[13] J. M. Madariaga, *Doctoral Thesis*, Universidad del Pais Vasco, Bilbao, Spain, 1983.
[14] A. M. Sastre, *Doctoral Thesis*, Universidad Autonoma de Barcelona, Barcelona, Spain, 1983.
[15] M. P. Elizalde, J. M. Castresana and M. Aguilar, *ISEC 83, International Solvent Extraction Conference, Proceedings*, Denver, USA, 1983; p. 491.
[16] L. A. Fernandez, M. P. Elizalde, J. M. Castresana, M. Aguilar and S. Wingefors, , *Solv. Extr. Ion Exch.*, 1985, **3**, 807.

[17] A. M. Sastre, N. Miralles, I. Casas and M. Aguilar, *Proc. Colloque Extraction Par Solvant et Echange d'Ions*, Toulouse, France, 1985.
[18] M. P. Elizalde, J. M. Castresana, M. Aguilar and M. Cox, *Solv. Extr. Ion Exch.*, 1985, **3**, 251.
[19] J. M. Castesana, M. P. Elizalde, M. Aguilar and M. Cox, *Chem. Scripta*, 1987, **27**, 273.
[20] M. E. Keeney and K. Osseo-Asare, *ISEC 83, International Solvent Extraction Conference, Proceedings*, Denver, USA, 1983; p. 345.
[21] D. S. Flett and S. Titmuss, *J. Inorg. Nucl. Chem.*, 1969, **31**, 2612.
[22] M. Cox and D. S. Flett, *ISEC 71, International Solvent Extraction Conference, Proceedings*, Denver, USA, 1983; p. 345.
[23] M. Aguilar, *Chem. Scripta*, 1973, **4**, 207.
[24] M. Aguilar and M. Muhammed, *J. Inorg. Nucl. Chem*, 1976, **6**, 1193.
[25] J. de Pablo, M. Aguilar and M. Valiente, *Chem. Scripta*, 1984, **24**, 147.
[26] L. G. Sillén, *Acta Chem. Scand.*, 1956, **10**, 186.
[27] L. G. Sillén, *Acta Chem. Scand.*, 1956, **10**, 803.
[28] M. P. Elizalde, J. M. Castresana, M. Aguilar and M. Cox, *Chem. Scripta*, 1984, **24**, 44.
[29] A. P. Brady, H. Huff and J. W. McBain, *J. Phys. Colloid. Chem.*, 1951, **55**, 304.
[30] E. B. Sandell, *Colorimetric Determination of Traces of Metals*, 3rd Ed., Interscience, New York, 1959.
[31] IUPAC, *Spectrophotometric Data for Colorimetric Analysis*, Butterworths, London, 1963.
[32] A. Sastre, N. Miralles and M. Aguilar, *Chem. Scripta*, 1984, **24**, 44.
[33] M. P. Elizalde, J. M. Castresana, M. C. Alonso and M. Aguilar, *Polyhedron*, 1985, **4**, 2097.
[34] D. H. Liem, *Acta Chem Scand.*, 1971, **25**, 1521.

6

Designing liquid-extraction equipment

P. J. BAILES
Schools of Chemical Engineering, University of Bradford, Bradford, West Yorkshire BD7 1DP, UK.

6.1 CONTACTOR SELECTION

A necessary precursor to the detailed design of a contactor must be an appreciation of the advantages of specific types of contactor and of the factors in a process that may lead to one type being preferred to another. For example, the principle features of columns, mixer-settlers and centrifugal contactors are sufficiently different that it will usually be clear to the designer that one type is going to be more suitable than the others. Numerous factors have a bearing on the initial choice and decisions are not always based on purely technical considerations. The features most likely to receive considerations are:

tradition/historical development, prejudice
cost, scale, geographical location
reliability of design
system properties, $\Delta\rho$, γ, emulsification
number of stages
contact time, product/solvent degradation
temperature of operation
contaminants, deposition of solids
flexibility, accommodation of flow changes
space available
control
maintenance, equipment simplicity
safety, enclosed design for flammable environment

The initial selection will often be based on flow-rate or extraction kinetics. Thus, the largest columns may cope with 100m³/hr but large mixer-settlers can deal with six times this amount. Where the extraction involves a slow chemical reaction, mixer-settlers are favoured. Alternatively, the effects of chemical reactions causing degradation of solute or solvent are minimized in centrifugal or column contactors which have shorter residence times.

A very important element in contactor selection is the degree of confidence that can be expressed in the design. Full-scale stagewise equipment is inherently linked to the required number of equilibrium stages through a stage efficiency that can be estimated with reasonable accuracy or determined precisely in small-scale equipment employing the same residence time as for full-scale. In the case of differential equipment, small pilot rigs are scaled-up with much less certainty, primarily because of the effects of scale on drop size and axial-mixing coefficients. Once operating experience has been achieved with a medium-sized column, these factors can be quantified and they become less of a problem but, this problem explains why new processes are often first put into practice with mixer-settlers.

Solvent-extraction equipment ranges from the small and sophisticated (centrifuges in pharmaceuticals, nuclear industry and metals extraction), to the large and rugged (concrete and steel mixer-settlers in copper extraction). Despite the variety of equipment [1], consideration of the methods of design of mixer-settlers and countercurrent columns should be sufficient for most purposes.

6.2   COLUMN DESIGN

Design of liquid–liquid extraction columns may be considered in two parts:

 (i) choice of the height to obtain the required degree of extraction
(ii) choice of the cross-sectional area required to obtain the given throughout

Treated separately they result, in most cases, in a very approximate design suitable only for initial costing purposes. This is because there is a high degree of interaction between the factors that govern mass transfer and hydrodynamic performance. Some possible column designs are shown in Figs. 6.1–6.4.

The height required for a contactor depends primarily on the rate of solute transfer $Q = KA\Delta C_{\text{mean}}$. The overall mass-transfer coefficient (K) depends on the individual film mass-transfer coefficients for the continuous and dispersed phases, and is therefore strongly dependent on the respective phase velocities and drop size. For high rates of transfer, it is important that the total interfacial area for mass transfer (A) is high. This can be obtained with a high degree of subdivision of the dispersed phase to give small drops with low rise velocities, and hence high volumetric fractions (or hold-up) of dispersed phase. Unfortunately, a sacrifice in throughput occurs when the drop size is reduced. Also, owing to axial mixing, the mixing flow patterns exert a considerable influence on the mean concentration driving force (ΔC_{mean}). Thus the mass-transfer process is shown to be strongly dependent on column hydrodynamics.

Practical studies of hold-up, drop size, mass-transfer and mixing behaviour have advanced basic understanding, particularly with regard to the design and interpretation of pilot plant experiments. Substantial progress has been made on essential elements such as the prediction of hold-up, flooding and axial mixing.

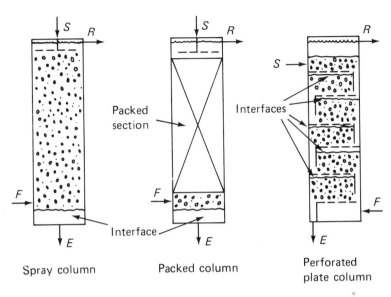

Fig. 6.1 — Three types of unagitated column contactors. Reprinted from *Chemical Engineering*, January 19, © 1976, by special permission of the copyright holders, McGraw-Hill, Inc., New York 10020.

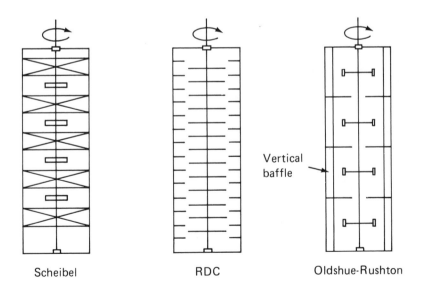

Fig. 6.2 — Three types of mechanically agitated columns. Reprinted from *Chemical Engineering*, January 19, © 1976, by special permission of the copyright holders, McGraw-Hill, Inc., New York 10020.

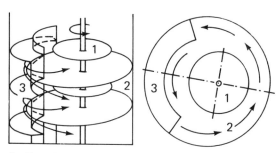

1. Rotating disc rotor
2. Mixing zone
3. Settling zone

Fig. 6.3 — The asymmetric rotating-disk (ARD) extractor. Reprinted from *Chemical Engineering*, January 19, © 1976, by special permission of the copyright holders, McGraw-Hill, Inc., New York 10020.

6.2.1 Hold-up
A slip velocity for counter-current flow of the two phases in a column without packing may be defined as follows:

$$V_s = \frac{V_d}{h} + \frac{V_c}{1-h}$$

where h is the fractional volumetric hold-up of dispersed phase, and the superficial velocities of dispersed and continuous phases are denoted by V_d and V_c, respectively. A helpful concept is that the slip velocity under the limiting condition of essentially zero hold-up is a characteristic velocity $\bar{V}_0$ bearing some relation to the velocity of a single drop moving under the constraints imposed by the internal design of the column. Attempts to determine the characteristic velocity theoretically from the terminal velocity of a single drop have had limited success but it does correlate quite well with the slip velocity. The simple empirical relation given by Thornton [2]

$$V_s = \bar{V}_0 (1-h)$$

works remarkably well for low values of hold-up. Other equations exist, notably that of Misek [3] which is of the form:

$$V_s = U_0 (1-h) \exp(Bh)$$

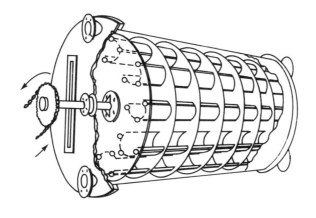

Fig. 6.4 — The Graesser raining-bucket contactor. Reprinted from *Chemical Engineering*, January 19, © 1976, by special permission of the copyright holders, McGraw-Hill, Inc., New York 10020.

where B is included as a measure of drop coalescence in the system.

Successful correlation between $\overline{V}_0$ and system properties, column geometry, extent of agitation etc., has been demonstrated for a variety of column types, but there are few reports on the application of such correlations to industrial columns.

6.2.2 Throughput
The maximum attainable throughput in a countercurrent column is determined either by phase inversion or entrainment. As the relative velocity of the phases

increases there must ultimately come a point at which the smallest dispersed phase drops are just held stationary. At higher flow-rates, increasing amounts of dispersed phase are carried out with the continuous phase and flooding occurs. In practice, this may happen only after prolonged operation and is not always immediately apparent.

The high throughput region of operation is one of increasing hold-up as shown in Fig. 6.5. Under such circumstances the flow limit may prove to be phase inversion rather than flooding by entrainment.

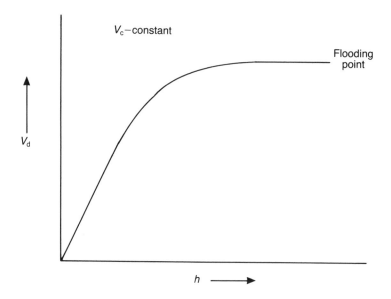

Fig. 6.5 — Hold-up behaviour as flooding is approached in a column.

Empirical flooding correlations are available for different columns, but they take no account of the direction of solute transfer. More valuable to the designer is the fact that the trend indicated in Fig. 6.5 suggests that differentiation of the relationship between slip velocity and hold-up might allow prediction of the flow-rates at flooding.

The Thornton equation may be differentiated to give superficial velocities at flooding:

$$V_{cf} = \overline{V}_0 (1 - 2 h_f) (1 - h_f)^2$$
$$V_{df} = 2\overline{V}_0 h_f^2 (1 - h_f)$$

Eliminating $\overline{V}_0$ between the expressions gives an equation for the hold-up at flooding. The assumption in the differentiation that $\overline{V}_0$ is constant up to and including the flooding region seems questionable. Certainly there is doubt as to whether the existing slip velocity equations pertain near flooding. Nevertheless, this approach does mean that experimental values of $\overline{V}_0$ from pilot plant can be used to give predictions of the full-scale column diameter at flooding.

6.2.3 Drop size

In any practical situation, a distribution of drop sizes exists as a result of coalescence and redispersion within the column. The absolute size of drops and the form of the size distribution can be strongly influenced by the direction of solute transfer, which affects drop-drop coalescence. Drop size affects both column height and diameter. The size of the drops controls the transfer rate since it affects the interfacial area available for transfer, turbulence levels within and around the individual drops and also the throughput of both phases. In general, the gain in area which results from making drops as small as possible more than compensates for the diffusional deficiences of such drops in terms of the overall rate of mass transfer. The limit to droplet subdivision is not imposed by the need to optimize between turbulence and interfacial area, but by the sacrifice in throughput which occurs when the drops are too small.

The use of pilot plant to determine characteristic velocities, number of transfer uinits and axial mixing coefficients eliminates the need to know drop size for scale-up. Also, mechanically agitated columns have sufficient flexibility to allow design deficiencies to be corrected by changing the rotor speed.

Column design from fundamental principles, however, depends on a much better understanding than currently exists of such things as drop break-up mechanisms, the relationship between characteristic velocity and drop size, and how this is affected by internal circulation within the drops.

6.2.4 Mass transfer and axial mixing

The traditional approach to the design of columns is to use the simple transfer-unit concept. This assumes countercurrent plug flow through the column such that the concentration driving force is maximized. However, it is usual in columns to get flow patterns that cause considerable deviations from plug flow.

All phenomena that lead to a distribution of residence times in the two phases are known collectively as axial mixing. The effect on ΔC is illustrated in Fig. 6.6. Axial mixing markedly lowers the mean concentration driving force for mass transfer, and therefore a column designed ignoring its presence will be too short for the required duty. Clearly, it is desirable to minimize the likelihood of such mixing occurring, but in many types of column it cannot be avoided and allowance must be made for its presence.

Mathematical models for axial mixing effects are available [4] (Fig. 6.7), but they require values for the dispersion coefficients. In the absence of pilot-scale tracer studies there are correlations for a variety of column types. Few of these refer to equipment bigger than 150 mm in diameter. Nearly all are concerned, however, with continuous-phase axial mixing only, since this tends to be the dominant effect.

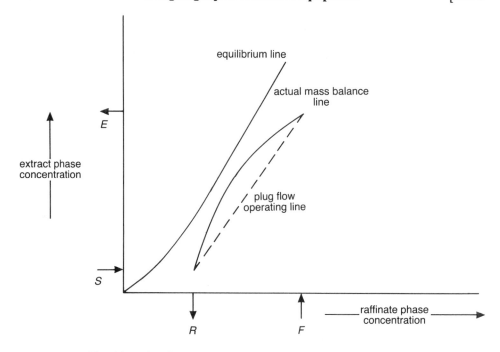

Fig. 6.6 — The effect of axial mixing on the concentration driving force.

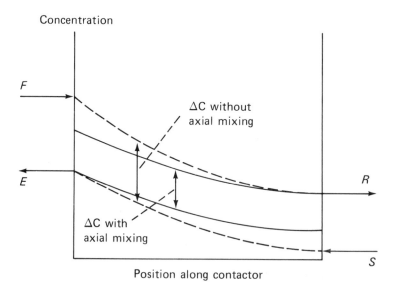

Fig. 6.7 — Effect of axial mixing on concentration profiles within column, and hence on value of ΔC. Reprinted from *Chemical Engineering*, January 19, © 1976, by special permission of the copyright holders, McGraw-Hill, Inc., New York 10020.

6.3 MIXER-SETTLERS

6.3.1 Stage equipment

Examples of mixer-settlers are shown in Figs. 6.8–6.10. The problems of axial mixing

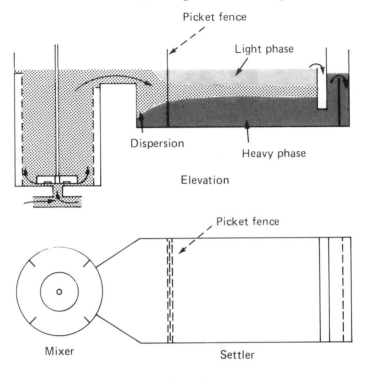

Fig. 6.8 — The General Mills mixer settler. Reprinted from *Chemical Engineering*, January 19, © 1976, by special permission of the copyright holders, McGraw-Hill, Inc., New York 10020.

found with differential equipment do not arise where there is stagewise contact of the two phases such as occurs in a series of mixer-settlers. This is because a stagewise contactor provides a number of discrete stages in which the two phases are brought to equilibrium (or near), separated, and passed countercurrent to the adjoining stages. It does not depend on the maintenance of an ideal concentration profile throughout the contactor. Mixer-settlers can be used, therefore, to reproduce almost exactly the results of classical stepwise calculations between operating and equilibrium lines. The reduction in risk that this brings to the design is offset by the necessity of separating the phases between each stage, which adds significantly to the overall size and liquid inventory.

6.3.2 The mixer

The sequence of events leading to the full-scale design of a mixer often commences with simple batch experimentation in a tank of geometry similar to that intended for the plant. Extraction kinetics can be examined in such equipment, and the fractional

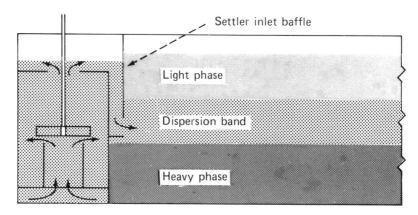

Fig. 6.9 — Main features of 'classic' Davy Powergas mixer-settler. (Not drawn to scale.)
Reprinted from *Chemical Engineering*, January 19, © 1976, by special permission of the
copyright holders, McGraw-Hill, Inc., New York 10020.

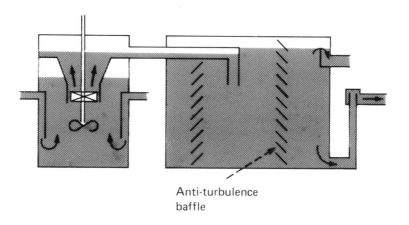

Fig. 6.10 — Essential features of I.M.I. mixer-settler. Reprinted from *Chemical Engineering*,
January 19, © 1976, by special permission of the copyright holders, McGraw-Hill, Inc., New
York 10020.

approach to equilibrium E_b determined as a function of time. This tends to be a first
order relationship which can be described by a rate constant k. The same rate
constant is then used to predict the extraction performance E_f of a continuous mixer

of the same size. The equations are analogous to those for a first-order reaction in batch and continuously stirred tank reactors thus:

$$E_b = 1 - \exp(-kt)$$

and

$$E_f = \frac{kt}{1 + kt}$$

The value of k is dependent on the impeller:tank ratio, impeller speed and tank geometry.

Small-scale continuous-flow tests with real process liquors are usually undertaken to determine the actual number of stages and also to examine the kinetics on a flow mixer. Scale-up from this point is almost always based on a fixed residence time known to give an acceptable approach to equilibrium in conjunction with constant tip speed or constant power per unit volume. The choice of procedure depends on the liquids and mixer design being used. Constant power per unit volume (P/V) tends to give high levels of power input at full scale and therefore high entrainment levels. Constant tip speed (πND) gives too low a power input and problems with mixer phase ratio, phase inversion and entrainment.

As a rule it is desirable to maintain tubulent conditions in the mixer. For baffled tanks the Power number ($P/\rho_c N^3 D^5$) under such circumstances is constant and provides a means of predicting power consumption. The influence of vortex formation at full scale may cause larger Power numbers to apply than those expected from the usual Power number–Reynolds number relationship based on experiments at a smaller scale [5].

In some designs the impeller provides for both pumping and mixing requirements. The clearance between a shrouded turbine impeller and the draught tube supplying the liquids can be used to control the degree of mixing in relation to pumping. In any event, power used for pumping is a small proportion of the total. It is important that sufficient mixing energy is supplied to avoid bypassing or stagnant zones within the mixer which would result in a different phase-ratio in the dispersion to that fed to the mixer. Curiously, a new development known as the "Combined Mixer Settler" [6] deliberately exploits the fact that the phase ratio in the dispersion need not correspond to the solvent:feed ratio. In this case, however, the situation is engineered so that the phase ratio in the dispersion is near to unity for good mixing, even though the solvent:feed ratio may be 10:1. With a mixer-settler this situation may be resolved by recycling one phase within a stage to achieve an optimum phase ratio different from the overall flow-ratio. This procedure, of course, increases the load on the settler.

Recognition of the similarity that exists between mixers for solvent extraction and continuous stirred tank reactors has led to the serious consideration of mixers in cascade as a means of reducing total mixer volume. In particular, smaller mixers allow savings in motor drives and gearboxes since these may then become standard, readily available sizes, rather than special units.

6.3.3 The settler

In the study of settler performance it is possible to establish a relationship between dispersion band thickness (Z) and dispersed-phase flow-rate per unit horizontal area of settler (Q_d/A). In many cases this may be of the form:

$$Z = a\left(\frac{Q_d}{A}\right)^y$$

also

$$\frac{Q_d}{A} = \frac{k_1 Z}{1 + k_2 Z}$$

Pilot-scale development work is advisable and the above equations provide a sound basis for this. It is important that the dispersion band should be of even depth and cover the whole settling area.

The principle worry for the designer is that there may be excessive entrainment of fine drops in the bulk phases leaving the settler. Entrainment problems may originate in the mixer owing to excessive tip speeds or vortex formation. More usually, however, they are a product of the coalescence process in the settler. Even distribution of the dispersion across the settler width and introduction at the level of the band at mid-depth of the settler are desirable features which reduce entrainment. The horizontal velocity in the two separated layers must be such as to discourage large-scale recirculation patterns sweeping fine drops from the dispersion band, but not so great that it distorts the dispersion band itself.

Correlations for dispersion-band thickness must be treated with caution since it is known that the physical properties of density, viscosity and interfacial tension are insufficient to correlate single-drop coalescence data. The presence of surfactants, extractant molecules and solid particles are all known to have a dramatic effect on coalescence performance.

6.4 CHOICE OF DISPERSED PHASE

The foregoing discussion on the design of columns and mixer-settlers presupposes that a decision has been made regarding which phase is to be dispersed. In fact, this choice can have a substantial bearing on equipment performance and may be influenced by a number of considerations. Factors that require evaluation are:

extraction rate and the possibility that there is a controlling resistance in one phase

the influence that direction of mass transfer may have on interfacial turbulence and hence extraction rate, and also its effect on coalescence.

differences in settling behaviour and entrainment

stability of the desired dispersion against inversion.

Imposed on the designer will be other constraints that may dictate the choice, such as: solvent:feed ratio; cost and/or safety measures aimed at minimizing solvent inventory; equipment characteristics, including the influence of the wetting properties of column internals and the ease of adjustment to the desired optimum phase ratio in mixer-settlers.

REFERENCES

[1] P. J. Bailes, C. Hanson and M. A. Hughes, *Chem. Eng.*, 1976, (January 19) 86.
[2] J. D. Thornton, *Chem. Eng. Sci.* 1956, **5**, 201.
[3] T. Misek, *Collect. Czech. Chem. Commun.*, 1963, **28**, 1631.
[4] S. Hartland, *Counter-Current Extraction*, Pergamon Press, Oxford, 1970.
[5] P. J. Bailes, J. C. Godfrey and M. J. Slater, *The Chem. Engr.*, 1981 (July) 331.
[6] J. B. Scuffham, *The Chem. Engr.*, 1981, (July) 328.

7

Liquid–liquid extraction in continuous flow analysis

M. Valcárcel
Department of Analytical Chemistry, University of Córdoba, Spain

7.1   INTRODUCTION

The growing demands of society in fields such as medicine, biology, ecology, industry, etc. have brought about a real revolution in analytical chemistry over the past few years. New analytical techniques and methods have been developed in response to the new problems posed. Amongst these, automatic methods of analysis have gained a notable momentum and have originated the development of a large variety of commercially available instruments.

There are three major types of automatic methods of analysis [1–3]: (a) discrete or batch methods, in which the sample maintains its identity in a reaction cup and the detection is discontinuous; (b) continuous methods, which involve the aspiration or injection of the sample into a reagent or carrier stream with continuous detection in a flow-cell. Various types can be recognized according to whether air-bubbles are employed (SFA) [4,5] or not [UFA] to avoid carryover between samples and, in the latter case, whether the sample has been injected (FIA) [3,6] or not (CCFA); and (c) robotic methods, which mimic the actions of an operator and which are still in the early stages of development [7].

In the endeavours to improve the automatic analytical methodology during recent years an important aspect has generally been neglected: sample pretreatment. Indeed, work on automatic methods of analysis has been almost exclusively devoted to aspects related to determination. Only in a few cases have the low sensitivity and selectivity of the original manual procedures been taken into account in automation. There is therefore a need for fully-automated handling systems [8].

Sample pre- or post-treatments are chiefly intended to: (1) improve selectivity, by removing interfering substances from the sample matrix, (2) improve sensitivity by means of preconcentration methods, (3) prevent the deterioration of the analytical system and (4) permit the analytical determination.

Several continuous-separation techniques have been incorporated in automation to achieve these objectives. For example, a membrane dialyser is a device commonly used in continuous segmented flow methods in clinical chemistry to remove proteins before the analytical reaction. Other separation techniques such as ion-exchange have been used less frequently. Liquid–liquid extraction is by far the most popular separation method for the clean-up and preconcentration of samples, because it is simple, reproducible, and versatile.

Liquid–liquid extraction methodology can be classified into two general types:- batch and continuous. Batch extraction methods are done in conventional funnels in one or more steps. Continuous liquid–liquid extraction methods can be performed in several ways: by solvent circulation, using a distillation–condensation apparatus [9], with the Craig apparatus [9,10] or by use of chromatographic techniques such as "extraction chromatography" [11] and "countercurrent chromatography" [12] which have great separating power.

The introduction of extractors in continuous (automatic) methods (SFA, FIA, CCFA) and pre- or post-column devices in HPLC are very promising and attractive for this field.

7.2 COMPONENTS OF A CONTINUOUS EXTRACTOR

The characteristics and working of each of the fundamental parts which make up the assembly of a continuous liquid–liquid extractor are as follows.

7.2.1 General scheme
A continuous liquid–liquid extraction system consists of three main parts (Fig. 7.1)

Fig. 7.1 — General scheme of a continuous liquid–liquid extractor.

and performs at least the following three functions:

(a) it receives the two streams of immiscible phases and originates a single flow with alternate and regular zones of the two phases. The part serving this dual purpose is called the "Solvent Segmenter".

(b) it facilitates the transfer of matter through the interfaces of the segmented flow in the "Extraction Coil", the length of which, together with the flow rate, determines the duration of the actual liquid–liquid extraction.

(c) it splits up, in a continuous manner, the segmented flow from the extraction coil into two flows of separate phases in a part called the "Phase Separator".

The working of these three parts is based on the same fundamental principle, i.e. the selective wetting of internal component surfaces by both organic and aqueous phases. In general, organic solvents wet Teflon surfaces whereas aqueous phases wet glass surfaces.

7.2.2   The solvent segmenter

This element acts as a confluence or mixing point of the two initial streams of the two phases which converge in it. Its main purpose is to obtain alternate and regular segments of the two immiscible liquids which enter the extraction coil. It must be designed so that the "length" of the segments can be checked conveniently. The commonest type is a simple modification of the Technicon Company connectors A-8 and A-10 (Fig. 7.2). These consist of a tube with three openings [13]. The aqueous

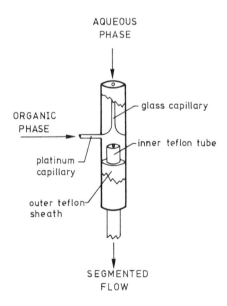

Fig. 7.2 — Adjustable solvent segmenter.

phase enters through a glass capillary and the organic one enters through a platinum capillary perpendicular to the former. Two adjustable concentric Teflon tubes shut off the tube at the other opening in the same direction as the aqueous phase. The height of the inner tube can be adjusted to that of the perpendicular opening. The length of the segments coming out of the main conduit of this tube depends on (a) the aforementioned height, (b) the inner volume of the mixing chamber and (c) the flow-rates of the two phases or their ratio (q_o/q_a), which is equivalent to the volume ratio in a discontinuous liquid–liquid extraction (which, in general, is approximately unity). The most common situation is to have identical segments, of about 5 mm in length.

There is also a possibility of carrying out this segmentation–mixing process by means of simple confluence points. Kawase [14] has studied the influence of three configurations with capillaries of an inner diameter of 0.8 mm: 90–90 T, 30–30 Y and 45–45 W which are shown in Fig. 7.3. With extraction coils of a larger inner diameter

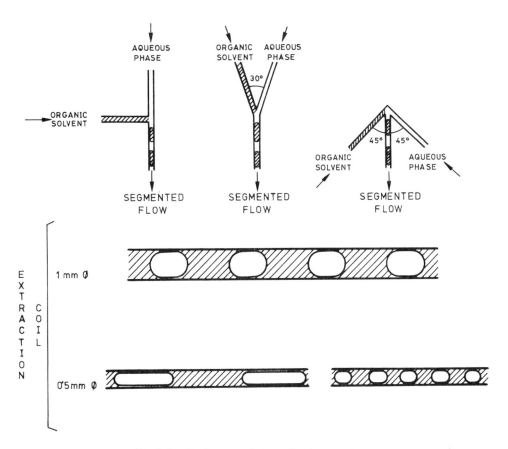

Fig. 7.3 — Confluence points as solvent segmenters.

(1 mm) the three configurations yield identical results, but for smaller diameters (0.5 mm) T and Y configurations give larger segments than W configuration.

The size of the segment may not affect extraction efficiency in a fast extraction but it does so in a slower system. However, the reproducibility of the segmentation is an important aspect. Fluctuation in solvent ratios caused by inconsistent segmentation makes the phase separator device give a faulty performance.

7.2.3 The extraction coil

This receives and holds for a specific time the segmented flow of the two phases and the transfer of matter between the phases takes place in it. Figure 7.4 depicts two

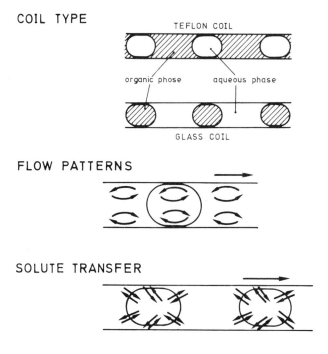

Fig. 7.4 — Characteristics of the extraction coil.

situations that occur, according to the type of coil. When this is made of Teflon the organic phase wets the walls and the aqueous phase is in the form of bubbles. If the coil is made of glass the situation is just the opposite. The flow patterns in coiled solvent segmented systems and the flow of solutes between the interfaces are also shown in Fig. 7.4. In addition to gravitational mixing, effective mass transfer results from plug flow in the opposite direction occurring in the segments, and also from the secondary flow patterns formed in a coiled tube. Therefore, tubes of a small diameter (0.5–1 mm) are recommended. Air-segmentation (in SFA) probably results in more active conditions thus making extraction more efficient.

Rossi *et al.* [15] recommend that the sample should always start off in the bubble phase. For instance, if the sample is in the aqueous phase, a variety of reasons make the use of a Teflon coil advisable. Sample carryover is then limited and the transfer speed of the solutes is increased. The extraction kinetics are improved if the analyte is allowed to reach the phase boundary easily since the distances inside the bubble are shorter. Moreover, the ratio of the interface area to the solute (sample) volume is higher when the solute is initially contained in the bubble phase.

The length of the coil will affect the extraction efficiency, and also the dispersion or dilution of the solute. The coil can be made long enough so that the transfer kinetics are not the limiting factor in extraction efficiency. Any increase in coil length beyond this point will result in increased sample dilution without a concurrent increase in sample recovery and is therefore undesirable.

In most cases the efficiency of the extraction is between 70 and 90%. It is therefore greatly influenced by the variables of the system, especially by the ratio of the flow-rates of the organic and the aqueous phases, which should be strictly controlled in order to obtain reproducible results.

The use of auxiliary processes such as the incorporation of beads in the coil, subjecting the coil to high temperatures, vibrations, ultrasonics, etc in order to improve its efficiency is not usually advisable as these cause emulsions which appreciably disrupt the working of the phase separator.

7.2.4 The phase separator

This receives the segmented flow from the coil and continuously splits it up into two separate flows. The separation is seldom quantitative: the efficiency usually ranges between 80 and 90%. However, the phase in which the determination is to be carried out should be kept completely free from the other, in order that on passing through the detector no spurious signals are generated.

The design of the phase separator must take into account the fact that the inner volume should be as small as possible in order to avoid parasitic/dispersion phenomena which, in addition to diluting the sample, thus yielding a weaker signal, result in loss of reproducibility.

In these continuous separator units the ratio of the flow rate of the stream coming into the solvent segmenter (q) and that going out of the phase separator (q') must be accurately regulated in the phase which finally contains the analyte or its reaction product. Control of the value of the ratio q/q' is vital in order to keep the efficiency of the process constant and obtain reproducible results. To achieve this objective it is essential to monitor the incoming stream, which is accomplished by making the waste tube go through the peristaltic pump beyond the cell.

There are three general types of continuous separators (Fig. 7.5):

(1) Devices using a chamber relying on gravity to separate the phases. The flow is separated continuously giving rise to a single interface. The denser phase comes out of the bottom part, the less dense out of the top. Figure 7.5 shows two possibilities as described by Bergamin [16] and Burguera [17].

(2) Devices with a "tee" separator using gravity with or without a phase guide made of a material wetted by one phase but not by the other [13]. Figure 7.5 shows a separator of this type which has a piece of Teflon attached to the tube interior, to help the separation of the less dense organic phase. Strips of hydrophobic paper can also be used for this purpose.

(3) Devices with a membrane phase separator based on the selective permeability of a microporous membrane (0.7–0.9 μm pore diameter) towards the phase which wets the membrane material, usually Teflon. The organic phase which goes through it is free from aqueous phase. Various configurations have been designed [14,18–21], but there are no significant differences between them. The

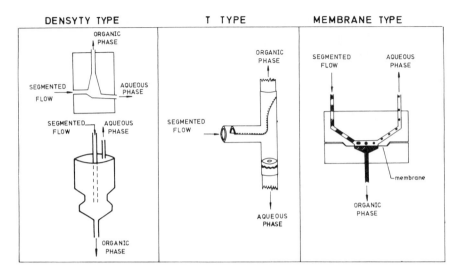

Fig. 7.5 — Different types of continuous phase-separators.

arrangement of the incoming (segmented) and outgoing streams (organic and aqueous phases) depends on the relative density of the phases.

The membrane phase separator has the following advantages over other types of continuous separators: (a) it has a smaller internal volume which lessens the dispersion or dilution of the analyte or its reaction product, which results in less band broadening and better sensitivity; (b) it is more reliable for separating phases at higher flow-rates so its use makes it possible to reduce the analysis time; (c) it can be used with a greater variety of water-immiscible solvents since a density difference between the aqueous and organic phases is not required; and (d) the separation efficiency is greater than for other separator devices (above 90–95%).

It is worth pointing out that in certain analyses, when a continuous fluorimetric detector is used, the method can be applied without employing a phase-separating device [2].

In the SFA mode, the phase separator receives a triply segmented flow consisting of air, organic phase and aqueous phase. Here, the less dense solvent comes out of the top exit together with the air bubbles and is contaminated by the other phase. Figure 7.6 depicts the segmented flow and two separators made by Technicon for solvents denser and less dense than the aqueous phase, respectively. Since the organic phase is the one which collects the analyte (or its reaction product) an internal Teflon insert helps to accomplish the separation. In these devices it is necesary to reincorporate the separating air bubbles at the exit of the organic phase to prevent carryover between samples [4,5].

7.3   CONTINUOUS (AUTOMATIC) FLOW METHODS

The main characteristics of a continuous liquid–liquid extraction of the sample, coupled on-line with SFA, FIA and CCFA systems, will be described next. Discrete

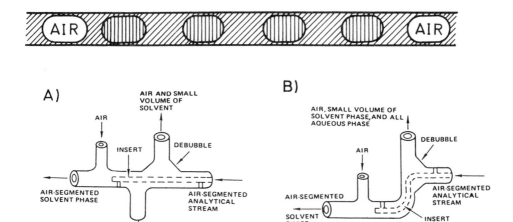

Fig. 7.6 — Phase separators in Segmented Flow Analysis (SFA). (Organic solvent denser (A) and less dense (B) than the aqueous phase).

automated liquid–liquid extraction sample-handing systems have not been considered because of their small importance. Solid-extraction semi-automatic systems (clean-up procedures based on adsorption, ion-exhange and gel permeation) [23–27] will not be dealt with either, as they show clear disadvantages as regards the stability and reproducibility of the phase system compared with liquid–liquid extraction.

7.3.1 Types of manifolds

Figure 7.7 shows a diagram of the manifolds most frequently employed in continuous automatic liquid–liquid extraction. All have the following features in common: a sampler, a peristaltic pump, two streams (one of a carrier, possibly containing a reagent, and another of organic solvent), a solvent segmenter, an extraction coil and a phase separator. The difference between the SFA and FIA configurations are: (a) the air bubbles inserted at two points before the carrier confluence and after the separation of the phases, and (b) the manner of inserting the samples (aspirated in SFA and injected in FIA). In the third configuration the sample is aspirated through a 2-way valve and there are three possibilities according to the type of detection system used: (a) CCFA, when the analyte phase flows directly to the flow-cell of a detector (b) FIA, when the analyte phase fills the loop of an injection valve and then the phase plug is carried to the detector unit and (c) discontinuous detection; the analyte phase is collected in several cups and detection is carried out individually.

Some of the most important applications of these general configurations will now be described.

7.3.2 Segmented-flow analysis

SFA methods were the first automatic ones to be developed by the Technicon company from a scheme by Skeegs. However, in recent years these have gradually

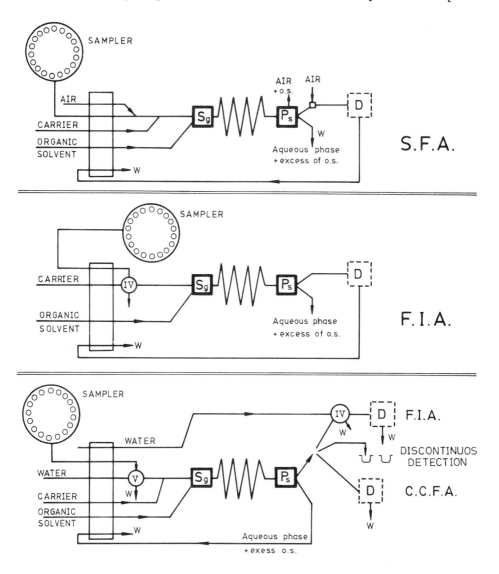

Fig. 7.7 — Continuous (automatic) extraction manifolds. SFA: air-segmented flow analysis. FIA: Flow-injection analysis. CCFA: Completely continuous flow analysis. (S_g: solvent segmenter, P_s: phase separator, IV: injection valve, D: detector, W: waste, o.s.: organic solvent).

been replaced by discrete methods and by the new continuous-flow (automatic) modes: FIA and CCFA.

A broad view of SFA liquid–liquid extraction configurations and their applications has been given in a monograph [4]. They are of historical value and their design and working have served as a basis for the development of liquid–liquid extractors in other modes of continuous-flow analysis.

The SFA manifold described by Carter and Nickless [28], shown in Fig. 7.8, has been selected as a typical example.

It was designed for the determination of metal ions by extraction of diethyldithio-carbamates previously formed in the aqueous phase of the sample in carbon tetrachloride. The absorbance of the organic phase is measured continuously. In this manifold a special phase-separator is used. The top of this is open to the atmosphere; the segmented flow goes in at the top through tube of a smaller diameter than that of the separator opening. The organic layer is pumped off from the bottom of the separator and the aqueous phase flows to waste through a side arm. This manifold is also an example of the use of a displacement flask system to prevent the organic solvent from passing directly through the pump tubes; an aqueous carrier displaces CCl_4 in closed bottles.

7.3.3 Flow-injection analysis
The continuous liquid–liquid extraction carried out in various FIA systems offers many advantages over SFA manifolds, such as: (a) lower sample, reagent and organic solvent consumption, (b) faster determinations, (c) simpler assemblies, (d) greater reproducibility, and (e) lower cost.

The first FIA methods associated with ligand–ligand extraction were proposed by Kalberg et al. [29] and Bergamin et al. [16] in 1978. In our recent monograph [3] we reviewed this subject, and described the different manifolds used and their applications. The most significant and representative cases will be described here.

Similarly to the manifolds described in Fig. 7.7, there are two general alternatives depending on whether injection takes place before or after the continuous extracter device.

The most common situation is when a prior injection of the sample has taken place. Figure 7.9a depicts a manifold for the determination of Vitamin B_1 (at the 3×10^{-4}–$6 \times 10^{-4} M$ level) in pharmaceuticals [30], based on the oxidation of thiamine to thiochrome in a carrier of potassium ferricyanide in a basic medium (NaOH). The thiochrome is continuously extracted in a chloroform stream. The fluorescence of the organic phase is measured continuously. A sampling rate of 70 samples per hour is easily achieved.

It is interesting to note that continuous multi-extraction is possible. Shelly and co-workers [13,15] have described such a system for the separation and determination of carcinogenic polynuclear aromatic compounds (pNAs) (Fig. 7.9b). A crude-oil-ash sample residue (500 μl) dissolved in cyclohexane is injected into a cyclohexane stream that merges with a DMSO stream in the first solvent segmenter. After passing through a glass extraction coil, the DMSO phase is mixed with a water stream in a cooling coil. In the second solvent segmenter a fresh stream of cyclohexane is segmented with the aqueous DMSO phase. After going through a Teflon extraction coil the organic phase is carried through the flow-cell of a video-fluorimeter. The optimized system has a sample recovery similar to an identical manual procedure and a 1.5% relative standard deviation between injections. The sampling frequency is 12 hr^{-1}.

Figure 7.9c shows a FIA assembly in which injection takes place after the continuous extractor. It was designed by ourselves [31] by adapting a manifold previously proposed by Karlberg et al. [32] for the determination of traces of

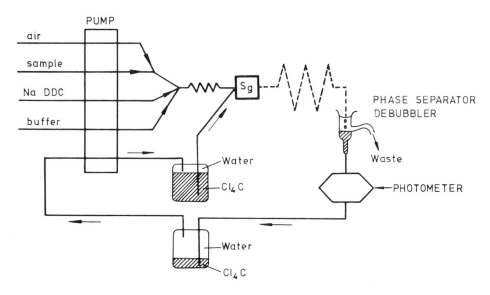

Fig. 7.8 — Air-segmented manifold for continuous extraction of metal ions with sodium diethyldithiocarbamate (NaDDCA) into CCl₄, using a displacement flask system.

perchlorate in serum and urine by formation of an ion pair with a cuproine-type chelate $(L_2Cu)^+$. The organic solvent (MIBK) contains the dissolved ligand, 6-methylpicolinealdehyde azine. The sample is continuously added to the system to be mixed with another stream containing Cu(II) and ascorbic acid. This mixture merges with the organic solution of the ligand in the solvent segmenter. The organic phase stream containing the ion pair, coming out of a 'T''-type separator, fills the 130-μl loop of an injection valve. This volume is injected into a water carrier which takes it to the atomic absorption spectrophotometer (AAS) in which the content of copper is determined continuously and hence that of perchlorate (0.5–5.0 μg/ml) is indirectly determined in serum and urine samples. The sampling frequency is 45 hr^{-1}. It is worth noting the reduction of the interferences caused by other bulky anions (i.e. I^-, NO_3^-, etc) in the manual method. This can be attributed to the shortening of the extraction time in the Teflon coil, which is to say that equilibrium has not been reached before the phase separator, and the interference of other anions is minimized since in the manual method the contact between phases is much longer and has a more pronounced effect.

7.3.4   Completely continuous flow analysis
This type of manifold is characterized by the lack of air bubbles and the absence of an injection system. The sample is continuously aspirated and the phase which finally contains the analyte goes continuously through the flow-cell of a detector. Only a few manifolds of this type have been devised. This configuration, sketched in Fig. 7.7, is a precursor of the extracting devices without any segmentation associated with HPLC.

Recently, Bengtsson *et al.* [33] and Backstrom *et al.* [34] have described, inde-

(a)

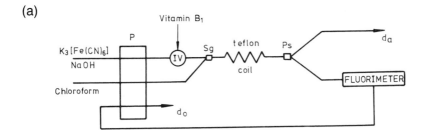

(b)

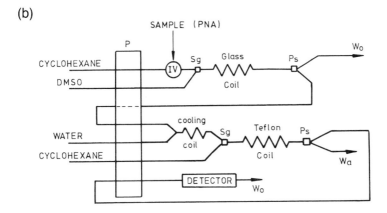

(c)

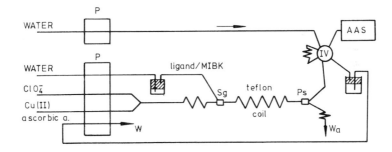

Fig. 7.9 — Typical FIA extraction manifolds. Injection takes place in front of the extractor components: simple (a) and multiextraction (b). Injection takes place after the extraction has been carried out (c) (based on systems described in [30,16,31]).

pendently, continuous liquid–liquid extraction systems for pre-concentration and clean-up of metal species in two stages: the extraction of metal carbonates in Freon 113 (1,1,2-chloro-1,2,1-trifluoroethane) and their subsequent re-extraction into an aqueous carrier containing Hg(II). The resulting aqueous phase is collected in sample cups, and detection is by graphite-furnace atomic-absorption spectrometry.

Therefore, these systems are not completely continuous since detection is carried out in a sequential manner.

7.4 CONTINUOUS EXTRACTION ASSOCIATED WITH HPLC

The introduction of continuous extractors coupled on-line with HPLC systems clearly improves some chromatographic determinations applied to agriculture, clinical chemistry and pharmacy. Although derivative-forming techniques in gas (GC) and liquid (LC) chromatography have been widely developed over the last 20 years, no continuous liquid–liquid extraction system associated with HPLC was described until 1978. It is interesting to note that no assembly of this type applied to GC can be found in the literature [35–37].

The most significant progress in this interesting area has been made mainly by two research groups, one led by Frei in Amsterdam and another one by Karger in the U.S.A.

A general distinction is made according to the situation of the continuous extraction system with respect to the chromatographic column.

7.4.1 Pre-column devices

The extraction system is located so that the solutes pass through it before reaching the HPLC column. Few examples of this assembly have been described. Other separation techniques, mainly adsoprtion, have been used coupled on-line with HPLC in order to achieve the same objectives, viz. clean-up and trace enrichment [38,39].

In 1978 Dolan *et al.* [36] reported an automatic analyser for fat-soluble vitamins (A_1, D_2 and E) in pharmaceutical tablets with an SFA-module located prior to an HPLC system with a UV-detector (254 nm). A schematic diagram of this analyser is given in Fig. 7.10. The overall CFA/HPLC system consists of (a) an automated sample preparation unit (Solid-prep II from Technicon) which receives the tablet from the sample tray, homogenizes it and places it in an ethanol–water mixture and (b) an air-segmented continuous extractor system that aspirates an aliquot from the sample and in which the vitamins are partitioned into hexane to increase the concentration by about 4.5 times. A stream of 5% sodium chloride is placed in the centre of the extraction coil to remove water-soluble interferences. After passing through the phase separator unit, the debubbled sample is drawn into a sample-loop for automated valve injection into an HPLC system. A microprocessor controls the automated solid unit, the injection valve and the UV-detector. Injection occurs 7.5 min after the start of the homogenization cycle.

This arrangement constitutes an excellent example of how the association of CFA and HPLC allows the solution of analytical problems that would be difficult to resolve by using the techniques separately. Thus the CFA module yields higher analysis rates and reduces the sample size compared with the corresponding manual methods, which require several hours and need 5–20 tablets for each analysis. The HPLC module facilitates the simultaneous detrmination because of its great resolving power and rapid separating ability; it is suitable for the analysis of all three vitamins (A_1, D_2 and E) in each sample, even though the amount of vitamin D_2 is lower by 1000-fold than that of the other two vitamins. Moreover, decomposition

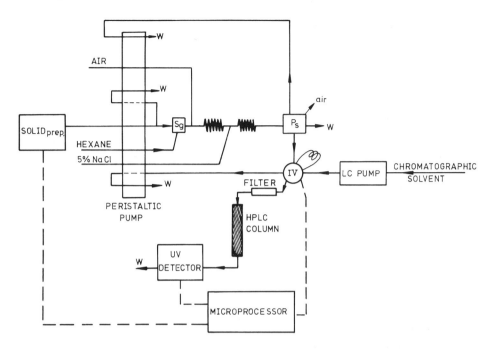

Fig. 7.10 — Air-segmented automatic analyser with a continuous extractor device prior to an HPLC system designed for the determination of fat-soluble vitamins (A_1, D_2 and E) in pharmaceutical tablets (based on a system described in [36]).

products from vitamin A acetate or E acetates, and components from the complex tablet matrix itself, tend to interfere with the analysis, particularly in the case of vitamin D_2.

7.4.2 Post-column devices
The use of post-column reactor–detectors in HPLC has gained widespread acceptance in recent years [40,41]. There are two general types:

(a) *Non-segmented*, in which the reagent merges directly with the effluent in several reactor models such as capillary [42] or bed [43] types which are useful for relatively fast reactions lasting for from a few seconds to several minutes.
(b) *Segmented* with air-bubbles (type 1) using the Technicon Autoanalyzer, or with immiscible organic solvents (type 2). The first type is used for slower reactions (5 min) to avoid excessive band broadening [44]. The second type of devices based on solvent segmentation, known as "*extraction detectors*", may use air-bubbles (SFA) or not (CCFA). They are employed to achieve one or more of the following objectives:

(1) to concentrate the analyte of the eluate, thus reducing band-broadening.
(2) to permit an otherwise impossible analytical detection when (a) the excess of

reagent interferes by giving the same signal as the reaction product, or (b) it is necessary to change the solvent for special detectors (e.g. a mass spectrometer).

Several detector configurations for fluorogenic ion-pair extraction, possibly involving air-segmentation, are shown in Fig. 7.11. These have been proposed by Frei and co-workers for the selective and sensitive determination of some basic drugs, pesticides and the metabolites. The ion-pairing reagent 9,10-dimethoxyanth-racene-2-sulphonate (DAS) is added once the reversed-phase chromatographic separation has taken place. The high-density organic extractant is then added and, after extraction of the ion-pair into the organic phase in the coils, the layers are separated in a conventional phase-separator. The ion-pair is then detected by its fluorescence, in the organic solvent. In the three-phase segmentation system (SFA) air-segmentation is used to reduce band broadening [35,45].

In a further step, an ion-pair extraction detector without air-segmentation (CCFA) for the determination of tertiary amines, drugs, chloro- and bromophenira-mine with DAS was reported [46] and secondary amines of pharmaceutical interest such as histidine, chloroxamine and fluoxamine with dansyl chloride have also been determined [47]. The "solvent segmentation" is a simple and effective alternative to three-phase systems because band broadening is kept below 20% with standard Technicon equipment and high reproducibility is achieved thanks to the relatively complex three-phase system which makes the phase-separator unit a vital part of the extraction detector. Tsuji has described an HPLC system with an ion-pair detector for the determination of erythromycin and erythromycin ethylsuccinate in serum with naphthotriazoledisulphonate as ion-pairing agent [37].

Recently a new and simpler configuration of ion-pair extraction detector has been proposed, involving the addition of the counter-ion agent (DAS) to the mobile phase prior to the column, for the determination of hydroxyatrazine in urine [48] and secoverine in biological samples [49]. The chromatographic effluent merges directly with the organic solvent stream in the solvent segmenter. The results obtained so far are very promising.

As a significant proportion of HPLC work is conducted under reversed-phase conditions, the direct coupling of a ligand chromatograph to a mass spectrometer is a problem which remains to be solved. Arpino and Guischon [50] and White [51] have reviewed the on-line HPLC–MS combinations proposed up to now. The major limitation of these methods is their inability to effectively remove, before the sample reaches the MS, either the solvent or other non-volatile constituents, or both. Karger and co-workers have proposed an interface consisting of a continuous extractor device either with [52] or without [53] air-segmentation. Figure 7.12 shows a schematic diagram of an HPLC–MS system utilizing a continuous extraction inter-face in the absence of air segmentation. A moving belt receives the organic phase with the extracted ion-pairs. The solute spectrum is superimposed but is easily discerned from the spectrum of the counter-ion.

7.5 CHROMATOGRAPHIC ASPECTS

It is not easy to draw a clear-cut distribution between the continuous liquid–liquid extraction methods already described and chromatographic partition methods.

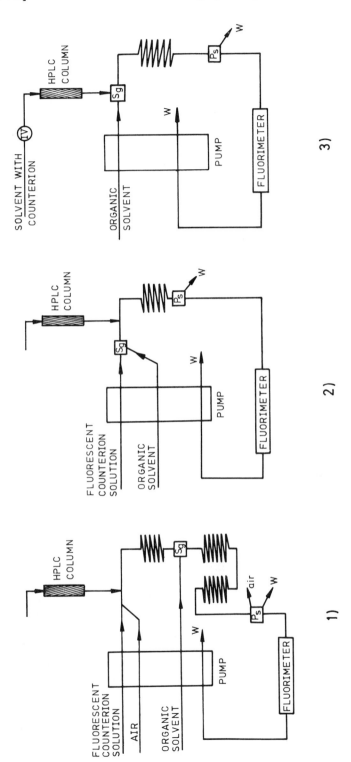

Fig. 7.11 — Fluorogenic ion-pair extraction detectors in HPLC (post-column reaction). 1. Three-phase system (SFA). 2. With solvent segmentation and post-column addition of reagent, 3. With solvent segmentation and pre-column addition of reagent.

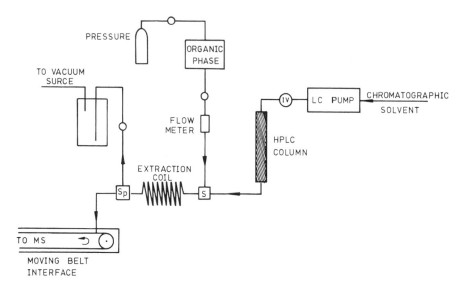

Fig. 7.12 — Schematic diagram of a reversed-phase HPLC–MS system with a non-air-segmented continuous extraction device and a moving-belt interface (based on a system described in [53]).

From a very general point of view (dynamic separation processes with a liquid–liquid interface) they can be considered jointly and therefore be included within the same group of analytical methods.

The Craig countercurrent extraction is an excellent example of continuous liquid–liquid extraction with an important relationship to chromatography. Its technical complexity makes it impracticable, although it is of great pedagogic interest in explaining the transition from static extraction to partition chromatography [9].

"Reversed-phase liquid column chromatography" is regarded by some authors as a synonym for "extraction chromatography". However, subtly different meanings are given to these terms by Braun and Ghersini [11]. These authors say that in normal reversed-phase partition chromatography the solute molecules undergo little, if any, chemical change (apart from association or proton exchange) but in extraction chromatography, a solute that is ionic initially is transferred from water (mobile phase) into an organic stationary phase, usually with complex chemical changes involving many interactions and equilibria.

In spite of all the analogies, there are major differences between extraction methods carried out in a continuous extractor, and chromatographic processes based on liquid–liquid partition. The most outstanding ones are as follows:

(a) The way in which the phases come into contact. In partition chromatography there is an inert support which retains the stationary phase, whereas both phrases are in motion in continuous extraction methods.
(b) The number of equilibrations (related to efficiency) is much greater in partition

chromatography as every chromatographic column works with a large number of theoretical plates.

(c) The analytical objective is different. In partition chromatography it is fundamentally directed towards the determination of several components in a single sample, whereas in continuous extraction methods it is used for pre-concentration or clean-up of many samples for the determination of a single component.

The recently developed countercurrent chromatographic methods (CCC) [12] are the best adaptation of liquid–liquid extraction to chromatography. They are characterized by the fact that they do not have a solid support to retain the stationary phase; they have advantages over other chromatographic methods in that they are free from all complications arising from the use of solid supports. The chromatographic column is replaced by a coil and and in this way (a) the stationary phase is retained in the open space of the coil (b) the coil acts as a separation column with hundreds and thousands of theoretical plates and (c) mass transfer resistance is minimized by providing a broad interface area and efficient mixing between the two phases. There are two basic continuous elution CCC systems, the hydrostatic equilibrium system (HSES), with a stationary coiled tube, and the hydrodynamic system (HDES) with a coil which is slowly rotated about its own axis. In both systems, after the mobile phase reaches the end of the coil it displaces only itself, leaving the stationary phase uniformly distributed along the column as segments of constant volume in each turn of the coil. Solutes introduced at the mobile-phase inlet are subjected to a partition process between the two phases. The hydrodynamic system yields much better resolution.

The performance of both CCC systems has been further improved by applying various patterns of centrifugal force fields with a rotary-seal-free-flow-through mechanism which permits continuous elution through multiple flow channels without risk of leakage or contamination. These systems provide a reliable elution system comparable to other chromatographic methods and are applicable to separation and purification of various biological meterials. A detailed description of these can be found in the review by Ito and Conway [12]. A commercial instrument has recently become available.

REFERENCES
[1] J. K. Foreman and P. B. Stockwell, *Automatic Chemical Analysis*, Ellis Horwood, Chichester, 1975.
[2] J. K. Foreman and P. B. Stockwell, *Topics in Automatic Chemical Analysis*, Vol. I, Ellis Horwood, Chichester, 1979.
[3] M. Valcarcel and M. D. Luque de Castro, *Análisis por Inyección en Flujo*, Dpto. Quimica Analitica y M.P.C.A. Córdoba (Spain), 1984; *Flow-Injection Analysis*, Ellis Horwood, Chichester, 1987.
[4] W. B. Furman, *Continuous Flow Analysis: Theory and Practice*, Dekker, New York, 1976.
[5] W. A. Coakley, *Handbook of Automatic Analysis. Continuous Flow Techniques*, Dekker, New York, 1982.
[6] J. Ruzicka and E. H. Hansen, *Flow-Injection Analysis*, Wiley, New York, 1981.
[7] J. N. Little, *Tr. Anal. Chem.*, 1983, **2**, 103.
[8] J. C. Kraak, *Tr. Anal. Chem.*, 1983, **2**, 183.
[9] M. Valcarcel and M. Silva, *Teoria y Práctica de la Extracción Liquido–liquido*, Ed. Alhambra, Madrid, 1984.
[10] L. C. Craig, *Anal. Chem.*, 1950, 22, 1346.
[11] T. Braun and G. Ghersini, *Extraction Chromatography*, Elsevier, Amsterdam, 1975.

[12] Y. Ito and W. D. Conway, *Anal. Chem.*, 1984, **56**, 534A.
[13] D. C. Shelly, T. M. Rossi and I. M. Warner, *Anal. Chem.*, 1982, **54**, 87.
[14] J. Kawase, J., *Anal. Chem.*, 1980, **52**, 2124.
[15] T. M. Rossi, D. C. Shelly and I. M. Warner, *Anal. Chem.*, 1982, **54**, 2056.
[16] F. H. Bergamin, J. X. Medeiros, B. F. Reis and E. A. G. Zagatto, *Anal. Chim. Acta*, 1978, **101**, 9.
[17] J. L. Burguera and M. Burguera, *Anal. Chim. Acta*, 1983, **153**, 207.
[18] L. Fossey and F. F. Cantwell, *Anal. Chem.*, 1983, **55**, 1882.
[19] K. Ogata, K. Taguchi and T. Imanad, *Anal. Chem.*, 1982, **54**, 2127.
[20] T. Imasaka, T. Harada and N. Ishibashi, *Anal. Chim. Acta*, 1981, **129**, 195.
[21] J. Kawase, A. Nakae and M. Yamanaka, *Anal. Chem.*, 1979, **51**, 1640.
[22] K. Kina, K. Shiraishi and N. Ishibashi, *Talanta*, 1978, **25**, 295.
[23] P. B. Stockwell, in *Methodological Survey*, Vol. 10, E. Reid, ed., Wiley, New York, 1980.
[24] E. Reid, *Merhodological Survey*, Vol. 7, Wiley, New York, 1978.
[25] R. C. Williams and J. L. Viola, *J. Chromatogr.*, 1979, **185**, 505.
[26] J. Lankelma and H. Poppe, *J. Chromatogr.*, 1978, **149**, 587.
[27] R. W. Frei and U. A. Brinkman, *Tr. Anal. Chem.*, 1981, **1**, 45.
[28] J. M. Carter and Nickless, G., *Analyst*, 1970, **95**, 148.
[29] B. Karlberg and S. Thelander, *Anal. Chim. Acta*, 1978, **98**, 1.
[30] B. Karlberg and S. Thelander, *Anal. Chim. Acta*, **114**, 129.
[31] M. Gallego and M. Valcarcel, *Anal. Chim. Acta*, in the press.
[32] L. Nord and B. Karlberg, *Anal. Chim. Acta*, 1983, **145**, 151.
[33] M. Bengtsson and G. Johansson, *Anal. Chim. Acta*, 1984, **158**, 147.
[34] K. Backstrom, L. G. Danielsson and L. Nord, *Analyst*, 1984, **109**, 323.
[35] J. C. Gfeller, G. Frey, J. M. Huen and J. P. Thevenin, *J. High Res. Chromatogr.*, 1978, **41**, 213.
[36] J. W. Dolan, J. R. Grant, N. Tanaka, R. W. Giese and B. L. Karger, *J. Chromatogr. Sci.*, 1978, **16**, 616.
[37] K. Tsuji, *J. Chromatogr.*, 1978, **158**, 337.
[38] HPM. Van Vliet, T. H. Bootsman, R. W. Frei and U. A. Brinkman, *J. Chromatogr.*, 1979, **185**, 483.
[39] S. K. Maitra, T. T. Yoshikawa, J. L. Hansen, I. Nilson-Ehle, W. J. Palin, M. C. Schotz and L. B. Guze, *Clin. Chem.*, 1977, **23**, 2275.
[40] J. F. Lawrence and R. W. Frei, *Chemical Derivatization in Liquid Chromatography*, Elsevier, Amsterdam, 1976.
[41] R. W. Frei and J. F. Lawrence (eds.) *Chemical Derivatization in Analytical Chemistry*, Vol. I (1981), Vol. II (1983), Academic Press, New York.
[42] R. W. Frei, L. Michel and W. J. Santi, *J. Chromatogr.*, 1977, **142**, 251.
[43] R. S. Deelder, M. G. F. Kroll, A. J. B. Beeren and J. H. M. van den Berg, *J. Chromatogr.*, 1978, **149**, 669.
[44] J. C. Gfeller, G. Frey and R. W. Frei, *J. Chromatogr.*, 1977, **142**, 271.
[45] R. W. Frei, J. F. Lawrence, U. A. Brinkman and I. L. Honingberg, *J. High. Res. Chromatogr.*, 1979, **2**, 11.
[46] J. F. Lawrence, U. A. T. Brinkman and R. W. Frei, *J. Chromatogr.*, 1979, **171**, 73.
[47] C. E. Werkhoven-Goewie, U. A. T. Brinkman and R. W. Frei, *Anal. Chim. Acta*, 1980, **114**, 147.
[48] C. van Buuren, J. F. Lawrence, U. A. T. Brinkman, I. L. Honingberg and R. W. Frei, *Anal. Chem.*, 1980, **52**, 700.
[49] R. J. Reddingius, G. H. De Jong, U. A. T. Brinkman and R. W. Frei, *J. Chromatogr.*, 1981, **205**, 77.
[50] P. J. Arpino and G. Guiochon, *Anal. Chem.*, 1979, **51**, 682A.
[51] P. C. White, *Analyst*, 1984, **109**, 973.
[52] B. L. Karger, D. P. Kirby, P. Vouros, R. L. Foltz and B. Hidy, *Anal. Chem.*, 1979, **51**, 2324.
[53] D. P. Kirby, P. Vouros, B. L. Karger, B. Hidy and C. Petersen, *J. Chromatogr.*, 1981, 203, 139.

8

New reagents

Michael Cox
Division of Chemistry, School of Natural Sciences, The Hatfield Polytechnic, Hatfield, Hertfordshire, U.K.

8.1 INTRODUCTION

The title 'New Reagents' is perhaps a little misleading, because over the past few years there has been little activity by the commercial reagent manufacturers to produce new reagents for liquid–liquid extraction. This is probably the result of the depressed state of the metals industry coupled with a reappraisal of the overall market for reagents. It should be realised that the total outlet for such speciality organic chemicals is relatively small and the insistence of the metal extraction industry and environmentalists that organic reagents should be recyclable is not advantageous for the reagent manufacturer. Thus it is not surprising that the early commercial reagents used for metal extraction were primarily produced for other applications. For example the amines were used in the detergency industry and the carboxylic acids in wood preservation, and it was only later that reagents were specifically designed for hydrometallurgical operations. However because this industry, like any other, is always looking for new and better products it is worthwhile considering what features are required for a successful commercial metal extraction reagent.

Organic reagents have been used for the selective concentration of metals in analytical chemistry for many years, and this use predates their commercial application in extractive metallurgy. Thus it is not surprising to find that many of the commercial reagents have similar molecular structures to the reagents used in analytical chemistry. However, the requirements for the two applications are quite different, so modifications are necessary. An organic metal extractant for hydrometallurgical use should have the following properties:

1. The reagent should be able to transfer a metal selectively across the aqueous/organic interface in both directions;
2. The reagent/diluent mixture should function efficiently with the proposed feed and stripping solutions in terms of rates of operation and stability towards degradation;
3. The reagent should offer maximum safety to plant and personnel, at minimum cost.

These properties are different from those for analytical applications. Thus, in analytical chemistry the metal need not be stripped from the organic phase. Also, the feed solution can be modified by the analyst, who can change the pH or add masking agents to improve the extraction without any restriction. Finally, because so little reagent is required, toxicity and cost are not prime considerations. Thus although general classes of organic molecule will find application in both analytical and industrial processes, the detailed structures of the compounds will be different. This can be seen by looking at examples in recent texts [1–3].

The above properties can be summarized in the following alliterative way [4]:

Strength   (the ability to transfer metal)
Selectivity
Speed   (kinetics of extraction and stripping)
Solubility   (compatibility of the reagent and the metal complex with the
                chosen diluent)
Stability   (resistance to degradation under the process conditions)
Separation   (ease of phase disengagement)
Sterling, $ (cost)
Safety   (toxicity, flammability, biodegradability)
Synthesis   (consistent reagent quality)
System   (good performance and interfacing with the overall hydrometallur-
                gical process)

These are not in any particular order and in some cases are inter-related. Thus the role of the chosen diluent is important not only for solubility but also for speed, separation, safety and system; indeed in considering a reagent for a hydrometallurgical process the test-work must also consider the likely diluent to be used.

In this paper, the main properties that will be discussed are strength and speed.

8.1.1 Strength

The strength of an extraction reagent relates to its ability to transfer metal across the aqueous/organic interface and depends on the formation of chemical bonds between the reagent and the metal ions to be transferred. These bonds can be ionic, with ion-pair reagents (1), or covalent with chelating acids [2] and solvating reagents [3].

$$2\,\overline{R_3NH^+} + MX_4^{2-} \rightleftharpoons \overline{(R_3NH^+)_2MX_4^{2-}} \tag{8.1}$$

$$2\,\overline{RCOCHC(OH)R} + M(H_2O)_x^{2+} \rightleftharpoons \overline{(RCOCHCOR)_2M} + 2H^+ \tag{8.2}$$

$$2\,\overline{R_2S} + MCl_2 \rightleftharpoons \overline{(R_2S)_2MCl_2} \tag{8.3}$$

where the bar denotes organic-phase species, and solvation of aqueous species is omitted.

The extraction chemistry of amines is covered in Chapter 4, so it is sufficient here to remember that extraction of metals occurs as a result of the co-ordination chemistry of the metal in the aqueous phase. The role of the amine is to accept a proton to form a cationic species; the tertiary amines, R_3N, have a high selectivity for the proton over metal ions.

The greatest developments in extractive metallurgy have arisen from the use of chelating acidic reagents. The ability to use concepts of co-ordination chemistry like the hard and soft acid base theory (HSAB) and formation (stability) constants to design selective reagents for metals has interested chemists for many years, and from these developments have emerged a number of successful commercial reagents.

As shown above (Eq. 8.2) the extraction process occurs by the replacement of an acidic proton on the reagent by a metal ion. The extraction is therefore sensitive to pH, a factor which provides another degree of freedom in the overall process. However this also produces an apparent anomaly if the extraction equation is written in the following way:

$$2\ \overline{RH} + M^{2+} + 2A^- \rightleftharpoons \overline{R_2M} + 2H^+ + 2A^-$$

where $A^- = Cl^-$, NO_3^-, $\frac{1}{2}SO_4^{2-}$, etc., or in general terms:

$$\overline{weak\ acid} + \frac{metal\ salt\ of}{strong\ acid} \rightleftharpoons \overline{\frac{metal\ salt\ of}{weak\ acid}} + strong\ acid$$

This equilibrium would be expected to lie well over to the left-hand side of the equation, so that little or no extraction would occur. However, it is well known that hydroxyoximes like LIX65N and P5100 extract copper from aqueous solutions at low pH, although they are very weak acids ($pK_a > 9$). This apparent anomaly can be explained in terms of co-ordination chemistry and the observation that ligand properties are altered by co-ordination to metals. Thus the extraction of copper by an aromatic hydroxyoxime can be represented by the following reaction scheme involving the 1:1 reagent:metal complex:

This scheme is then repeated with another reagent molecule to produce the extracted complex of the type R_2M. A similar scheme can be written involving the dimeric form of the reagent, which is known to exist under certain circumstances. This loss of the proton from the co-ordinated complex is facilitated because donation of electrons from the oxygen to the metal increases the acidity of the proton, as has been shown for other systems [5].

8.1.2 Speed

The rate of extraction depends on the concentration of species at the reaction site, which under normal hydrometallurgical conditions is the aqueous/organic interface. The observed surface activity arises from a balance between the hydrophilic parts of the molecule, as typified by the polar inorganic donor atoms, and the lipophilic organic substitutents. This interfacial activity is also dependent on the nature of the diluent and the bulk phase concentration of the reagent [7]. It is important that metal extraction reagents have a high interfacial activity, but this should not be too great, or problems will occur with phase disengagement (separation). In the past these properties were examined largely by trial and error, but with a better understanding of the interfacial chemistry involved it is now possible to provide guidelines for the choice of particular alkyl substituents to reagent molecules and for particular diluents for rapid kinetics [6].

The application of the ideas outlined above can be illustrtated by considering some reagents currently available.

8.2 HYDROXYOXIMES

Considerable development has occurred since the production of the first hydroxy-benzophenone oxime in 1965 by General Mills (now Henkel), designated LIX64. The 5-dodecyl alkyl substituent of this extractant was replaced in 1969 by a 5-nonyl chain (LIX65N) to give the major component in the very successful copper reagent LIX64N. The molecular structure of 2-hydroxy-5-nonylbenzophenone oxime was based on the analytical reagent cupron, benzoin oxime. However, the presence in LIX65N of the benzene ring close to the active donor atoms means that the interfacial area occupied by the molecule is quite large and this contributes to the relatively slow kinetics. Replacement of this aryl group by methyl (SME529, Shell) and finally by a proton (P50, P1, ICI) has, as expected, improved the kinetics of operation with subsequent benefits for plant operation [7]. In addition because the latter reagents are isomerically pure in containing only the active *anti* isomers, higher loading capacities are obtained than with LIX65N.

An interesting development by ICI/Acorga was the addition of nonyl phenol to the reagent P50 to produce the P5000 series of reagents. The reason for this deliberate addition was to reduce the extractive strength and improve the strip performance of P50 [8] and allow the compatability of the P5000 series with existing copper extraction plants. However, because of the deleterious effect of nonyl phenol on rubber linings, tridecanol has replaced it in the PT5050 reagent.

Although the aryl hydroxyoximes have been extensively studied over the past fifteen years both in industrial and academic laboratories, there are still interesting facets of their extraction chemistry which are poorly understood, one of which is the mechanism of the acceleration on addition of LIX63 in the mixed reagent LIX64N.

8.3 ORGANOPHOSPHORUS ACIDS

This group includes organic esters of phosphoric acids, $(RO)_2PO(OH)$; phosphonic acids, $(RO)RPO(OH)$ and phosphinic acids, $R_2PO(OH)$. Of these, the alkylphosphoric derivatives, especially di-2-ethylhexylphosphoric acid (D2EHPA), have proved to be the most versatile to date. As a reagent, D2EHPA has the desirable properties outlined above, and also the advantage of versatility for the extraction of many metals, for example, zinc, uranium, vanadium, cobalt, nickel, and rare earths. Its continued use since 1949 underlines its commercial success. It is only recently that alkyl derivatives of the phosphonic and phosphinic acids have become commercially available, although they were considered, along with D2EHPA, for rare-earth and actinide extraction in the 1950s.

The two commercial reagents are 2-ethylhexylphosphonic acid mono-2-ethylhexyl ester (PC-88A, Daihachi Chemicals, Japan) and di(2,4,4-trimethylpentyl)-phosphinic acid (Cyanex 272, Cyanamid Corpn.). There has also been a report [3] of the commercial availability of the di-2-ethylhexylphosphinic acid, which would be very interesting to compare with D2EHPA and PC-88A. Two aspects of the extraction chemistry of these organophosphorus acids are of interest, the behaviour of the alkaline-earth metals and the separation of cobalt and nickel.

One of the problems in the use of D2EHPA for the extraction of zinc is the co-extraction of calcium. With all simple acidic reagents the order of metal extraction closely follows the order of metal hydrolysis constants, with the metal ions of highest charge being extracted at the lowest pH. Where ions of the same charge exist, then extractability with D2EHPA seems to be related inversely to the ionic radii. Thus, the relative extractability follows the order Ba<Sr<Mg<Ca, since the ionic radii decrease in the order Ba>Sr>Ca>Mg. However although the order of extractibility of zinc, calcium and magnesium follow the same general order, there is some variation when the series of reagents phosphoric, phosphonic, and phosphinic acids are considered. This can be shown by comparing the pH_{50} values for these reagents determined under exactly the same experimental conditions and given in Table 8.1. Note the variation in the position of calcium; in the case of the reagent Cyanex

Table 8.1 — pH_{50} values of selected elements with organophosphorus extractants

	D2EHPA	PC-88A	DOPA*
	(0.1 M in MSB210)		(0.1 M in toluene)
zinc	1.66	1.9	2.15
calcium	2.25	3.2	4.80
copper	2.84	2.8	3.85
magnesium	3.36	4.0	5.30
cobalt	3.55	3.0	4.70
nickel	3.99	4.3	5.92

* di-n-octylphosphinic acid.

272 the position of calcium and magnesium is reversed. This could have interesting commercial applications and studies are continuing to try to elucidate this anomalous behaviour of calcium and investigate other similar systems [9].

The separation of cobalt and nickel has excited considerable interest in recent years. The easiest way to separate these two very similar elements is to exploit the fact the cobalt, unlike nickel, readily forms anionic chloro complexes. Thus a number of commercial processes are based on a chloride leach followed by amine extraction of the anionic cobalt complex [10]. Organophosphorus extractants can also be used to separate these elements from sulphate solutions. The interesting feature of this process, common to all these reagents, is that it is the chemistry of the organic phase which determines the separation and not the nature of aqueous phase species. Thus, as the cobalt concentration increases and the temperature of extraction increases the extracted cobalt complex changes from a hexa-co-ordinated partially hydrated species, $CoL_2(H_2O)_2$, to a polymeric anhydrous tetra-co-ordinated complex, $(CoL_2)_x$. The nickel complex of all these reagents, on the other hand, retains the hexa-co-ordinated form, $NiL_2(H_2O)_2$. The dependence of the separation factor on the amount of cobalt in the tetrahedral form has been demonstrated by Flett [10]; very large separation factors can be achieved by using these newer reagents. Thus under the same experimental conditions the cobalt/nickel separation factors are D2EHPA 14; PC-88A 280; Cyanex 272 7000.

8.4   SOLVATING REAGENTS

Like the chelating acid extractants, these provide the opportunity for selective processes by careful choice of the donor atom. Recent developments in this area have concentrated on providing reagents with sulphur donor atoms, to take advantage of its 'softness' and therefore selectivity for the precious metals. Examples are the dialkyl sulphides such as the alkylphosphine sulphide (Cyanex 471) which is reported as being selective for palladium and silver.

The most commonly used solvating reagent is TBP (tri-n-butyl phosphate). Developments of related reagents are still taking place and manufacturers, particularly Mobil and Daihachi, are producing alkyl phosphates $(RO)_3PO$, and alkylphosphonates $(RO)_2RPO$, for which applications are sought. One such development in zinc extraction will be discussed later (Chapter 11).

The final reagent to be considered is the latest extractant from ICI (DS5443) designed to extract copper from chloride media [4]. This is a solvating reagent that extracts copper according to the equation:

$$2\overline{L} + Cu^{2+} + 2Cl^- \rightleftharpoons \overline{CuL_2Cl_2}$$

Thus the reagent transfers copper without pH control, but the extraction is dependent on chloride ion concentration and although the precise structure is not known, it is believed to be an alkyl ester of a pyridine carboxylic acid [12].

Pyridine itself is a well known donor molecule for many metals and it can also act as a proton acceptor. However, this reagent is highly selective for metals over the proton; it is only 30% protonated in 10 M acid. The reagent can therefore extract copper from concentrated hydrochloric acid solutions as well as from solutions high in chloride concentration. Selectivity is very good; for example ferric iron extraction follows the protonation curve because iron is extracted as $FeCl_4^-$. Some other observations are less easy to explain; thus although pyridine co-ordinates readily to both copper and zinc, DS5443 has a high selectivity for copper over zinc, and the greater the electron withdrawing power of the carboxylic ester grouping the greater is the selectivity. Thus, in this new reagent there is scope for a lot of interesting extraction chemistry, but its introduction to commercial processing is conditional not on the solvent extraction step but on solving the problem of electrowinning of copper from strong chloride media. This is an illustration of the need to consider the whole hydrometallurgical process when designing new reagents.

8.5 CONCLUDING REMARKS

Undoubtedly the production of new reagents is decreasing. It is perhaps significant that the situation has remained static since 1977, with the only advances in commercial reagent manufacture being the alkylphosphonic and alkylphosphinic acids and the solvating reagents mentioned above. Thus, although selective reagents for nickel, zinc, etc., can be made, it is not certain whether the industry requires them or would be prepared to pay the economic price for a relatively small market requirement. On the other hand, with over 40 reagents commercially available and at least 12 in daily use, is there any need for future development? Surely it is not beyond the wit of chemists to adapt feed conditions to satisfy the requirements of existing reagents, or to use combinations of reagents in synergistic mixtures to alter the selectivity series associated with both reagents or increase the rate of mass transfer. Perhaps what is required is more and better data with the existing reagents under varying conditions. However it must be remembered that this situation is viewed in the context of the current world economy and existing technology. A change in either of these factors could alter the picture completely and a need for more reagents designed for a particular application may emerge. For example, the current work in liquid membranes, both emulsion and polymer-supported, uses reagents which are freely available commercially. Perhaps once the specific demands of this new technology are known, new reagents will be required.

REFERENCES

[1] T. Sekine and Y. Hasegawa, *Solvent Extraction Chemistry*, Dekker, New York, 1977.
[2] T. C. Lo, M. H. I. Baird and C. Hanson, eds., *Handbook of Solvent Extraction*, Wiley, New York, 1983.
[3] G. M. Ritcey and A. W. Ashbrook, *Solvent Extraction*, Part I, Elsevier, Amsterdam, 1984.
[4] R. F. Dalton, R. Price and P. M. Quan, *International Solvent Extraction Conference*, *Proceedings*, Denver, Colorado, 1983; p. 189.
[5] G. I. H. Hanania and D. H. Irving, *J. Chem. Soc.* **1962**, 2745.
[6] M. Cox, M. J. Grey, C. G., Hirons and N. Miralles, in *Hydrometallurgy, Research, Development and Plant Practice*, Osseo-Asare and Miller, eds., AIME, 1984; p. 357.
[7] J. F. C. Fisher and C. W. Notebaart, [2], Chapter 25.1.
[8] J. A. J. Tumilty, J. P. Massam and G. W. Seward, *International Solvent Extraction Conference*, *Proceedings*, Toronto 1977, volume 2, 1979; p. 542.
[9] J. Tatum and N. Miralles, unpublished results.
[10] G. M. Ritcey, [2], Chapter 25.2.

[11] D. S. Flett, in *Hydrometallurgy, Research, Development and Plant Practice*, Osseo-Asare and Miller, eds., AIME, 1984; p. 50.

[12] R. Price, *Nitrogen-Containing Extractants*, lecture to Solvent Extraction and Ion Exchange Group, Society of Chemical Industry, May 1984.

9

Supported liquid membranes†

Pier R. Danesi*
Chemistry Division, Argonne National Laboratory, 9700 South Cass Avenue, Argonne, Illinois 60439

The possibility of utilizing thin layers of organic solutions of solvent-extraction reagents, immobilized on microporous inert supports interposed between two aqueous solutions, for selectively removing metal ions from a mixture, represents an attractive alternative to liquid–liquid extraction. The permeation of metal species through such immobilized liquid layers can be described formally as the simultaneous combination in a single stage of an extraction and a stripping operation, occurring under non-equilibrium conditions. A detailed knowledge of liquid–liquid extraction equilibria and mass-transfer kinetics is therefore required to understand and to describe quantitatively the rate laws which control the permeation of metal species through supported liquid membranes (SLM), and to exploit them for separation processes. SLMs appear particularly promising for these kinds of processes in view of the following advantages over traditional separation technologies:

(1) lower capital and operating cost,
(2) low energy consumption,
(3) the possibility of using economically expensive, tailor-made extractant molecules,
(4) the possibility of achieving high separation factors.

The mechanism of coupled transport through SLMs is schematically described in Fig. 9.1. The SLM consists of a solution, in a water-immiscible low dielectric constant organic diluent, of an extracting reagent — a metal carrier in membrane terminology — absorbed in a microporous polymeric film having a thickness ranging from 25–50 μm. The polymeric film, which acts as a solid support for the liquid membrane, is generally made of polypropylene, polysulphone or other hydrophobic material, and

† Work performed under the auspices of the Office of Basic Energy Sciences, Division of Chemical Sciences, U.S. Department of Energy, under contract number W-31-109-ENG-38.
* Present address: International Atomic Energy Agency, Seibersdorf Laboratory, Vienna, Austria.

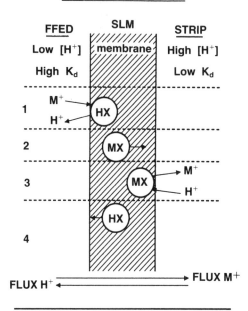

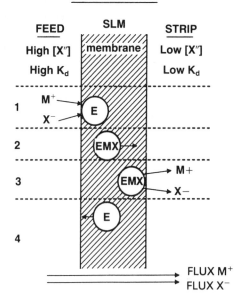

Fig. 9.1 — The mechanism of coupled transport through supported liquid membranes. Reprinted from P. R. Danesi, *Sep. Sci. Technol.,* 1984–5, **19**, 857, by permission of the copyright holders, Marcel Dekker Inc.

has pore sizes ranging from 0.02 to 1 μm. The solid support can be shaped as a flat sheet or as a tiny hollow fibre. In the second case the organic solution of the extracting reagent is absorbed in the microporous walls of the fibre (Fig. 9.2). The

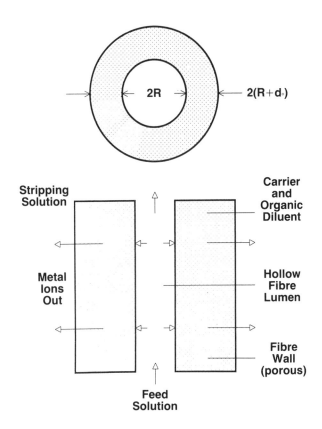

Fig. 9.2 — A hollow-fibre solid support. Reprinted from P. R. Danesi, *Sep. Sci. Technol.*, 1984–5, **19**, 857, by permission of the copyright holders, Marcel Dekker Inc.

SLM is interposed between two aqueous solutions. The aqueous solution that initially contains all the metal ioins which can permeate the SLM is referred to as the feed solution. The distribution ratio between the organic phase absorbed in the membrane pores and the aqueous feed solution of the metal species permeating the SLM, K_d, is here made high enough to favour metal extraction into the membrane phase. The aqueous solution present on the opposite side of the membrane, which is initially free from the permeable metal ions, is referred to as the strip solution. In this case the distribution ratio K_d is made as low as possible in order to favour complete back extraction of the metal species from the liquid membrane. If the metal carrier is an acidic extractant, HX, the difference in K_d between the feed and strip sides of the SLM is generally achieved by a pH gradient. This process is a counter-transport phenomenon (Fig. 9.1, upper part) and the chemical reaction which is responsible for the coupled-transport can be represented as:

$$M^+ + \overline{HX} \text{ (membrane)} \frac{\text{Feed Side}}{\text{Strip Side}} \overline{MX} \text{ (membrane)} + H^+ \qquad (9.1)$$

If the metal carrier is a neutral or a basic extractant (i.e. a long chain alkylamine), E, the difference in K_d between feed and strip is generally obtained by a concentration gradient of the counter-ion, X^-, which is accompanying the metal cation into the membrane. In this case, the process is a co-transport phenomenon (Fig. 9.1, lower part) and the chemical reaction which is responsible for the coupled transport can be represented as:

$$M^+ + X^- + \overline{E} \text{ (membrane)} \frac{\text{Feed Side}}{\text{Strip Side}} \overline{EMX} \text{ (membrane)} \qquad (9.2)$$

pH and counter-ion concentration gradients are most often used as driving forces.

From this schematic description of coupled transport it can be seen that metal species can be transported across the membrane against their concentration gradient. This "uphill" transport continues until all the metal species which can permeate the SLM have been transferred from the feed to the strip side, providing that the driving force of the process is constant. It follows that in an SLM permeation process, very high concentration factors can be obtained by using a volume of the strip solution which is much lower than that of the feed solution. Moreover, by using carrier molecules, HX or E, which are very selective to given metal species, very clean separation processes can be performed. Since during the permeation the carrier acts as a shuttle, moving metal species from the feed to the strip solution and then diffusing back (being continuously regenerated during the process) very small amounts of carrier are used in SLM separations. As a consequence, very expensive, highly selective, tailor-made carriers can be used economically. Other potential advantages of SLM separations over separations performed by traditional solvent-extraction techniques are the lack of solvent entrainment phenomena (leading to high separation factors), the possibility of handling feed solutions containing suspended solids, the simplicity of the equipment involved and the low energy consumption of the process. Moreover, compared to separation processes performed with solid membranes, SLMs offer the additional advantage of higher fluxes since diffusion in liquids is much faster than in solids.

The various steps which characterize the transport of metal species through SLMs can be described with the help of Fig. 9.1. *Step 1*: The metal species, after diffusing to the feed solution –SLM interface, react with the metal carrier. Protons are simultaneously released into the feed solution (counter-transport, acidic carrier) or X^- ions accompany the metal ions into the membrane (co-transport, neutral or basic carriers). *Step 2*: The metal–carrier complex diffuses across the membrane because its concentration gradient is negative. *Step 3*: At the SLM–strip solution interface the metal–carrier complex releases metal ions into the aqueous phase. Protons replace the M^+ ions in the membrane (counter-transport) or X^- ions are simultaneously released together with M^+ ions into the strip solution (co-transport). *Step 4*: The uncomplexed carrier diffuses back across the membrane.

When the distribution ratio at the membrane–aqueous strip interface is much lower than at the membrane–aqueous feed interface and the membrane phase polarity is low enough to make negligible the concentration of charged species with respect to that of uncharged ones, the steady-state overall membrane flux can be derived by applying the Fick's diffusion law to the aqueous boundary layer present on the feed side of the membrane, to the membrane itself and by expressing the inerfacial flux in terms of the interfacial kinetics. If the concentration of the metal-containing species is much lower than that of the carrier and of the H^+ or X^- ions in the feed solution, their concentration gradients are practically constant and the three equations which describe the fluxes are:

(a) $$J_a = -D_a \frac{\partial [M^+]}{\partial x}$$ (9.3)

(*diffusion through the aqueous boundary layer*), where D_a is the aqueous diffusion coefficient of the metal containing species,

(b) $$J_b = K_1 [M^+]_i - K_{-2} [\overline{M}]_i$$ (9.4)

(*interfacial flux*), where K_1 and K_{-1} are the pseudo-first-order rate constants of the interfacial reactions described in Eq. (9.1) or (9.2) and $[M^+]_i$ and $[\overline{M}]_i$ are the interfacial concentrations of the metal containing species at the feed solution–membrane interface on the aqueous and membrane side respectively,

(c) $$J_c = -D_o \frac{\partial [\overline{M}]}{\partial x}$$ (9.5)

(*membrane diffusion*), where D_o is the membrane diffusion coefficient of the metal-containing species $\overline{M}$. At the steady state, $J_a = J_b = J_c$ and by making a further assumption of linear concentration gradients, it can be shown that the following equation holds for the membrane flux, J, and the permeability coefficient P:

$$P = \frac{J}{C} = \frac{K_1}{K_1 \Delta_a + K_{-1} \Delta_o + 1}$$ (9.6)

The meaning of Δ_a and Δ_o and a schematic description of the processes involved are shown in Fig. 9.3. C is the time-dependent bulk concentration of the metal species in the feed solution. By dividing the numerator and the denominator of Eq. (9.6) by K_{-1} and considering that $K_1/K_{-1} = [\overline{M}]/[M^+] = K_d$, when the interfacial chemical reactions are very fast (local equilibrium), Eq. (9.6) simplifies to

$$P = \frac{J}{C} = \frac{K_d}{K_d \Delta_a + \Delta_o}$$ (9.7)

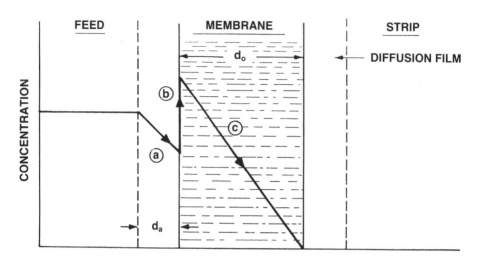

Fig. 9.3 — Diagram showing the membrane flux processes. Reprinted from P. R. Danesi, *Sep. Sci. Technol.*, 1984–5, **19**, 857, by permission of the copyright holders, Marcel Dekker Inc.

Since the relationship which relates the membrane flux to C, to the aqueous feed volume V and to the membrane area Q, is

$$J = -\frac{dC}{dt} \times \frac{V}{Q} \qquad\qquad (9.8)$$

the integrated form of the flux equation is

$$\ln \frac{C}{C_o} = -\frac{Q}{V} \times P \times t \qquad\qquad (9.9)$$

C_o is the value of C at time zero.

Equations (9.7) and (9.9) predict the permeation behaviour of SLMs when the feed solutions are relatively dilute in metal species. In this case, the permeability coefficient is a time-independent parameter containing the chemical and diffusional parameters characteristic of each metal ion permeating through a given SLM in contact with a given aqueous feed solution and the membrane flux J is varying with C. When the feed solutions contain metal ions at relatively high concentrations, and reactions (9.1) or (9.2) are completely shifted to the right, K_d is no longer independent of C. The highest value of $[\overline{M}]$ is $[\overline{HX}]/n$ (counter-transport) or $[\overline{E}]/n$ (co-transport) and the distribution ratio $K_d = [\overline{M}]/C$ becomes inversely dependent on C; n is the number of carrier molecules per metal ion in the metal–carrier complex and $[\overline{HX}]$ and $[\overline{E}]$ are the total carrier concentrations in the SLM. In this situation, for counter-transport, Eq. (9.7) becomes

$$\frac{J}{C} = \frac{([\overline{HX}]/nC)}{([\overline{HX}]/nC)\Delta_a + \Delta_o} \tag{9.10}$$

and for sufficiently large values of C

$$J = \frac{[\overline{HX}]}{n\Delta_o} \tag{9.11}$$

An identical equation holds for co-transport, if $[\overline{HX}]$ is replaced by $[\overline{E}]$. In order for Eq. (9.11) to hold, a further assumption that the membrane counter diffusion of the carrier is a very fast process with respect to the diffusion of the metal–carrier complex must be introduced. Equation (9.11) can be integrated to

$$C = C_o - \frac{[\overline{HX}]}{n\Delta_o} \frac{Q}{V} t \tag{9.12}$$

Equation (9.12) shows that the metal concentration in the feed solution decreases linearly with time. Therefore, the time-independent flux is a more useful parameter to describe the permeation of metal species through SLMs. Obviously during the permeation of the metal species, starting with relatively high metal concentrations, there is a transition region where the decrease of concentration with time is expressed neither by equation (9.12) nor by equation (9.9). In this region the flux equation cannot be analytically integrated and a numerical solution for the integrated flux equation, containing a concentration dependent P, has to be obtained.

SELECTED REFERENCES

[1] R. Marr and A. Kopp, *Int. Chem. Eng.*, 1982, **22**, 44.
[2] E. L. Cussler and D. F. Evans, *Sep. Purif. Methods*, 1974, **3**, 399.
[3] P. R. Danesi, *Sep. Sci. Technol.*, 1984–5, **19**, 857.
[4] P. R. Danesi, E. P. Horwitz, and P. G. Rickert, *J. Phys. Chem.*, 1983, **87**, 4709.
[5] P. R. Danesi, E. P. Horwitz, and G. F. Vandegrift, *Sep. Sci. Technol.*, 1981, **16**, 201.

10

Application of a simple model to the thermodynamics of liquid and solid cation exchangers

Erik Högfeldt
Department of Inorganic Chemistry, The Royal Institute of Technology, S-100 44 Stockholm, Sweden

During the past decade the liquid cation exchanger dinonylnaphthalenesulphonic acid, HD, has been studied in order to compare it with solid resins of the sulphonic acid type [1]. In the present paper some of the results obtained will be given in order to illustrate that the same model applies to liquid and solid cation exchangers with reference to ion-exchange and water uptake data.

10.1 THE MODEL

The basis for the model is given by Guggenheim's zeroth approximation [2a]. In a binary mixture of A and B, the number of A–A pairs, B–B pairs and A–B pairs are *proportional* to (not equal to, as suggested [3]) x_A^2, x_B^2 and $2x_A x_B$. Here x_A and x_B are the stoichiometric mole fractions of A and B.

When this model is applied to ion exchange, it is assumed that any extensive thermodynamic property, Y, is dependent on composition as follows

$$Y = y_A x_A^2 + y_B x_B^2 + 2y_m x_A x_B \tag{10.1}$$

In ion exchange, x_A and x_B are the equivalent fractions of the two ions in the exchanger phase, y_a and y_B are the quantity Y for the pure ionic forms and y_m refers to the mixed system. In any mixture there is a certain amount of each pure ionic form, and the distributions of the three kinds are assumed to be given by the random distribution of the zeroth approximation. The proportionality constants are incorporated in the values for y_A, y_B etc. Thus Eq. (10.1) is obtained whether it is assumed that the number of pairs is equal to or proportional to x_A^2, etc.

It is convenient to relate the homogeneous equation (10.1) to

$$Y = y_A x_A + y_B x_B + B x_A x_B \tag{10.2}$$

where

$$y_m = \tfrac{1}{2}(y_A + y_B + B) \tag{10.3}$$

and B is a quantity taking care of deviations from linearity (additivity) in the plot $Y(x)$.

For $B = 0$ or constant, y_m is a *constant* obtained from Eq. (10.3). For second degree polynomials in x the constants y_A, y_B and y_m have theoretical meaning; for higher-degree polynomials it becomes only a convenient way of reducing data to a few parameters.

In the following, the model will be applied to ion-exchange equilibria and water uptake during exchange.

10.1.1 Ion-exchange equilibria

For the sake of simplicity only uni-univalent exchange will be considered. For the ion-exchange reaction

$$A^+ + \overline{BX} \rightleftharpoons \overline{AX} + B^+ \quad (X^- = \text{ion-exchange group}) \tag{10.4}$$

the equilibrium quotient, κ, is given by

$$\kappa = \frac{[AX][B^+]}{[BX][A^+]} (y_B/y_A) \tag{10.5}$$

where y_A and y_B are the activity coefficients of the two cations in the aqueous phase. In the following it is assumed that their ratio stays practically constant and can be included in κ.

According to Eq. (10.2) the function $\log \kappa(x)$ can be expressed by

$$\log \kappa = \log \kappa(1)x + \log \kappa(0)\,(1-x) + Bx(1-x) \tag{10.6}$$

where $x =$ equivalent fraction of AX.

According to Eq. (10.3)

$$\log \kappa_m = \tfrac{1}{2}\{\log \kappa(0) + \log \kappa(1) + B\} \tag{10.7}$$

In ion exchange, the limiting values are not experimentally available by direct experiment. Sometimes, however, tracer studies can be done at concentrations low enough for good estimates of $\kappa(0)$ and $\kappa(1)$ to be obtained [4]. The integral free energy of reaction (10.4) can be transformed into a thermodynamic equilibrium constant, K, by [5–7]

$$\log K = \int_0^1 \log \kappa(x) dx \qquad (10.8)$$

From Eqs. (10.6)–(10.8)

$$\log K = \tfrac{1}{3}(\log \kappa(0) + \log \kappa(1) + \log \kappa_m) \qquad (10.9)$$

If we write Eq. (10.6) as a polynomial

$$\log \kappa = a + bx + cx^2$$

the following expressions are obtained for the activity coefficients f_{AX} and f_{BX} in the exchanger phase

$$\log f_{AX} = (1-x)^2 \, (\tfrac{1}{2}b + \tfrac{1}{3}c + \tfrac{2}{3}cx) \qquad (10.10a,b)$$
$$\log f_{BX} = x^2(\tfrac{1}{2}b + \tfrac{2}{3}cx)$$

where $a = \log \kappa(0)$

$$b = 2\,\{\log \kappa_m - \log \kappa(0)\} \qquad (10.11a–c)$$

$$c = \log \kappa(0) + \log \kappa(1) - 2\log \kappa_m$$

For $c = 0$ (i.e. $B = 0$), $\log \kappa(x)$ is a straight line and the expressions for the activity coefficients simplify to

$$\log f_{AX} = \tfrac{1}{2}b(1-x)^2$$
$$\log f_{BX} = \tfrac{1}{2}bx^2$$

i.e. the equations relating to regular solutions [2b]. Besides free energies, enthalpies and entropies can also be expressed by Eq. (10.2).

10.1.2 Water uptake
The number of water molecules per sulphonate group, W, is given by

$$W = ([H_2O]_{tot,org} - [H_2O]_{dil})/[HD] \qquad (10.12a)$$

for the liquid ion exchanger HD dissolved in a suitable diluent. The amount of water extracted by the diluent has to be subtracted from the total water content in the organic phase.

For solid resins, W is obtained from

$$W = n_{H_2O}/S_0 \qquad\qquad (10.12b)$$

where n_{H_2O} is the number of millimoles of water in a certain sample and S_0 the capacity of that sample expressed in milliequivalents. Observe that $w(0)$ and $w(1)$ are easily determined by measuring the water uptake by samples containing the pure ionic forms AX and BX.

10.2 EXAMPLES

The approach outlined above will be applied to both HD and sulphonic-acid type resins.

10.2.1 The system $CH_3NH_3^+–H^+$ on HD in heptane
(a) *Ion exchange equilbria.* Mikulich [8] studied the ion exchange between H^+ and a number of alkylammonium salts, using a $0.100M$ solution of HD in heptane and an ionic strength in aqueous phase of $0.100M$.

In Fig. 10.1 log κ is plotted against $x(=x_{CH_3NH_3D})$ for the temperatures 0°C,

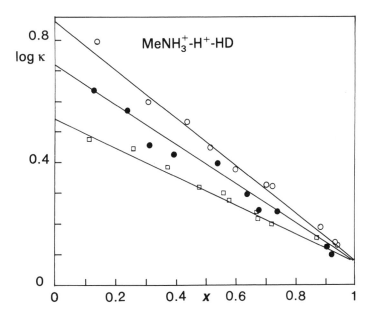

Fig. 10.1 — Log κ plotted against x for the system $CH_3NH_3^+–H^+$ on HD in heptane. Data from [8]. ○ 0°C; ● 25°C; □ 63°C. The curves were computed from the constants given in Table 10.1.

25°C and 63°C. The data have been fitted by three straight lines, i.e. $B = 0$, and the regular solution model applies to the activity coefficients.

By linear regression the constants log $\kappa(0)$ and log $\kappa(1)$ were obtained. These constants were fitted to expressions log $\kappa(1/T)$ giving

$$\log \kappa(0) = -0.861 + 471/T \tag{10.13a,b}$$

$$\log \kappa(1) = 0.079$$

The straight lines in Fig. 10.1 were computed by using Eqs. (10.13a,b). In Table 10.1 experimental and computed values for log κ are compared. Finally the

Table 10.1 — Comparison between experimental and computed log κ-values for the reaction:

$$CH_3NH_3^+ + HD \rightleftharpoons CH_3NH_3D + H^+$$

The reaction was studied at three temperatures: 0°C, 25°C and 63°C. $I = 0.100M$ (CH₃NH₃,H)Cl. The constants were computed from Eqs. (10.13a,b)

	0°C			25°C			63°C	
x	log κ exp	log κ calc	x	log κ exp	log κ calc	x	log κ exp	log κ calc
0	—	0.863	0	—	0.718	0	—	0.540
0.140	0.796	0.753	0.129	0.636	0.636	0.113	0.475	0.488
0.310	0.598	0.620	0.273	0.569	0.567	0.260	0.447	0.420
0.442	0.533	0.516	0.314	0.457	0.517	0.376	0.384	0.367
0.517	0.450	0.458	0.395	0.427	0.466	0.480	0.320	0.319
0.598	0.382	0.394	0.540	0.397	0.373	0.569	0.301	0.278
0.702	0.325	0.313	0.640	0.302	0.309	0.570	0.280	0.277
0.723	0.322	0.296	0.681	0.247	0.283	0.672	0.239	0.230
0.884	0.193	0.170	0.739	0.242	0.246	0.675	0.217	0.229
0.934	0.140	0.131	0.904	0.126	0.140	0.718	0.199	0.209
0.946	0.136	0.121	0.923	0.102	0.128	0.869	0.154	0.139
1	—	0.079	1	—	0.079	1	—	0.079
	σ(log κ)	±0.022		σ(log κ)	±0.030		σ(log κ)	±0.016

spread $\sigma(\log \kappa)$ is given in Table 10.1. This quantity is computed from

$$\sigma(\log \kappa) = \pm \sqrt{\Sigma(\log \kappa_{exp} - \log \kappa_{calc})^2 (n-1)} \tag{10.14}$$

where n = number of experimental points.

A fit of ± 0.02 to ± 0.03 is obtained, which is acceptable. With only three parameters, the equilibrium data for the system $CH_3NH_3^+ - H^+$ on HD can be fitted over the whole concentration range from pure ammonium salt to pure acid form, and a temperature range of about 60 K.

(b) *Water uptake.* In Fig. 10.2, W is plotted against x for the equilibrium data at 25°C in Fig. 10.1. Here, the data can not be fitted by a straight line, i.e. $B \neq 0$. By least-squares methods, the following expression was obtained.

$$W = 3.96x + 10.17(1-x) - 6.42x(1-x) \qquad \sigma(W) = \pm 0.23 \tag{10.15a}$$

If the values $w(0)$ and $w(1)$ are regarded as 100% certain, B can be computed for each experimental point giving

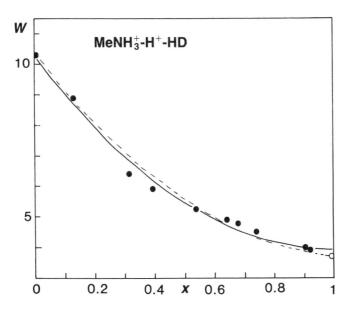

Fig. 10.2 — W plotted against x for the system $CH_3NH_3^+$–H^+ on HD in heptane at 25°C. Data from [8]. ———— Computed from Eq. (10.15a); – – – Computed from Eq. (10.15b).

$$W = 3.74x + 10.30(1-x) - 5.27x(1-x) \qquad \sigma(W) = \pm 0.27$$
$$(10.15b)$$

In Fig. 10.2, the curves corresponding to Eqs. (15a,b) are drawn: ———— Eq. (15a), – – – Eq. (15b). From Fig. 10.2 it is evident that Eq. (10.15b) with only one unknown parameter, B, gives a satisfactory fit to the data. From Eq. (10.3) the following value is obtained

$$w_m = 3.86 \ [\text{Eq. (10.15a)}]$$
$$w_m = 4.39 \ [\text{Eq. (10.15b)}]$$

The W-value for the mixed form is thus close to that for pure CH_3NH_3D.

10.2.2 The system $CH_3NH_3^+$–H^+ on KRS-8

(a) *Ion exchange equilibria*. Mikulich [8] also studied the same ammonium–hydrogen exchanges on the solid resin KRS-8 which is a polystyrenesulphonate with 8% crosslinking.

In Fig 10.3, log κ is plotted against x for this resin and the same three temperatures as before. Here, however, a second degree polynomial is needed to fit the data. By using linear regression on data log $\kappa_i(1/T)$ the following expressions were obtained

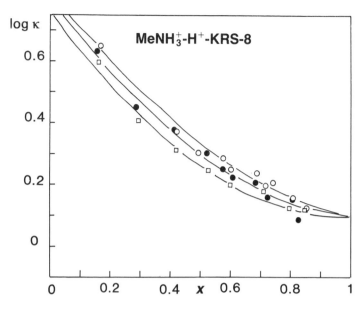

Fig. 10.3 — Log κ plotted against x for the system $CH_3NH_3^+-H^+$ on KRS-8. $I = 0.100M$ $(CH_3NH_3,H)Cl$. Data from [8]. ○ 0°C; ● 25°C; □ 63°C. The curves were computed from the model by using the constants given in table 10.2.

$$\log \kappa(0) = 0.528 + 86/T$$

$$\log \kappa(1) = 0.101 \qquad\qquad (10.16a\text{–}c)$$

$$\log \kappa_m = -0.446 + 175/T$$

The curves in Fig. 10.3 were computed with the constants given in Table 10.2,

Table 10. 2 — Constants in Eq. (10.6) computed from Eqs. (10.16a–c) for the system $CH_3NH_3^+-H^+$ on KRS-8

$T\,°C$	$T\,K$	$\log \kappa(0)$	$\log \kappa(1)$	$\log \kappa_m$	B	$\sigma(\log \kappa)$
0	273.2	0.843	0.101	0.195	−0.554	±0.018
25	298.2	0.816	0.101	0.141	−0.635	±0.023
63	336.2	0.784	0.101	0.075	−0.735	±0.016

obtained from Eqs. (10.16a–c) together with Eqs. (10.3) and (10.6). In Table 10.2 $\sigma(\log \kappa)$ computed from Eq. (10.14) is also given for each temperature. A comparison with Table 10.1 shows that about the same fit is obtained, but here five parameters are needed.

(b) *Water uptake*. In Fig. 10.4, W is plotted against x for the equilibrium data at 25°C

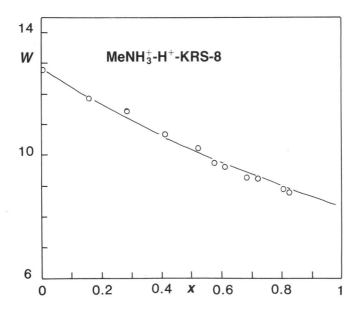

Fig. 10.4 — W plotted against x for the system $CH_3NH_3^+–H^+$ on KRS-8. $T = 298$ K. Data from
[8]. The curve was computed with the constants in set 3 in Table 10.3.

in Fig. 10.3. By least-squares methods the data were fitted to a straight line and a
second degree polynomial. Again, by assuming $w(0)$ and $w(1)$ to be known
constants, B has been computed for each experimental point giving set 3 in Table
10.3. The last column in Table 10.3 gives $\sigma(W)$. From Table 10.3 it is evident that
an acceptable fit is obtained with only one unknown parameter.

Table 10.3 — Parameters to fit Eq. (10.2) for $Y = W$. The system $CH_3NH_3^+–H^+$ on KRS-8 at 298 K

Set No.	Method	$w(0)$	$w(1)$	B	w_m	$\sigma(W)$
1	Linear regression	12.61	8.04	0	10.33	±0.14
2	Least squares	12.81	8.26	−1.26	9.91	±0.09
3	B from each point	12.76	8.34	−1.43	9.84	±0.10

10.2.3 The system $Na^+–H^+$ on Dowex 50 of various degrees of crosslinkage

Bonner and Rhett [9] studied the sodium–hydrogen exchange on Dowex 50 of three
different degrees of crosslinkage: 4, 8 and 16% DVB. In Fig. 10.5, log κ is plotted
against x_{NaR} for these data. The equilibrium quotient refers to the reaction

$$Na^+ + HR \rightleftharpoons NaR + H^+ \qquad (R^- = \text{resin anion})$$

By least-squares methods the data were fitted with second degree polynomials. The

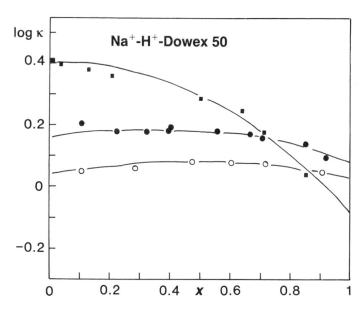

Fig. 10.5 — $\log \kappa$ plotted against x_{NaR} for the system Na^+–H^+ on Dowex 50 of various degrees of crosslinkage. $I = 0.100M$. $T = 298$ K. Data from [9]. $\bigcirc$ 4% DVB; $\bullet$ 8% DVB; $\blacksquare$ 16% DVB. The curves were computed from the model with the constants given in Table 10.4.

three constants were then studied as function of the degree of crosslinkage(%), giving

$$\log \kappa(0) = -0.080 + 0.030(\%)$$
$$\log \kappa(1) = -0.117 + 0.050(\%) - 0.003(\%)^2 \qquad (10.17\text{a–c})$$
$$\log \kappa_m = 0.026 + 0.026(\%)$$

In Table 10.4, the constants computed from Eqs. (10.17a–c) are given, together with

Table 10.4 — Constants computed from Eqs. (10.17a–c). System Na^+–H^+ on Dowex 50 of various degrees of crosslinkage. $T = 298$ K. $I = 0.100M$. Data from [9]

% DVB	$\log \kappa(0)$	$\log \kappa(1)$	$\log \kappa_m$	B	$\sigma(\log \kappa)$
4	0.040	0.035	0.130	0.185	± 0.009
8	0.160	0.091	0.234	0.217	± 0.013
16	0.400	-0.085	0.442	0.569	± 0.020

$\sigma(\log \kappa)$ for each degree of crosslinkage. The curves in Fig. 10.5 are those computed from the constants in Table 10.4. From Table 10.4 and Fig. 10.5 it is evident that expressions (10.17a–c) give a satisfactory fit to the experimental data. Thus, the model outlined above permits data reduction, not only of composition and temperature, but also of degree of crosslinkage for polystyrene sulphonates.

10.3 INFLUENCE OF SIZE ON THE ION-EXCHANGE PROCESS

HD and its salts form reverse micelles in heptane; Mikulich [10] measured the size of the aggregates by vapour phase osmometry (VPO). In Fig. 10.6, the number

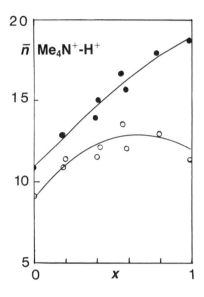

Fig. 10.6 — $\bar{n}$ plotted against $x_{(CH_3)_4ND}$ for the system $(CH_3)_4N^+$–H^+ on HD in heptane. $T = 298$ K. Data from [10]. ○ Dry samples; ● Wet samples. The wet samples were equilibrated with an aqueous phase.

average, $\bar{n}$, as estimated by the VPO method, is plotted against composition for the system $(CH_3)_4N^+$–H^+ on HD in heptane at 25°C. In spite of the fact that the size is approximately doubled during the ion-exchange process, the model outlined above applies to the data for log $\kappa(x)$. The conclusion is that the ion-exchange process is a site effect, independent of aggregate size for the micelles of HD and its salts. That is, they behave like ion-exchange resins, where each single bead can be regarded as a very large aggregate.

With long-chain amines and their salts, the conclusion is not so easy to draw. Here sometimes the data seem to be independent of size, at other times the size is important. This, however, is for the monomer–dimer equilibria in benzene studied by Aguilar and associates. The question now arises, how large must the aggregates be in order to behave as a separate phase (i.e. independent of aggregate size)? Future work might give an answer.

REFERENCES

[1] E. Högfeldt, R. Chiarizia, P. R. Danesi and V. S. Soldatov, *Chem. Scripta*, 1981, **18**, 13.
[2] E. A. Guggenheim, *Mixtures*, Clarendon Press, Oxford, 1952, (a) p. 30, (b) p.34.
[3] E. Högfeldt, *Reactive Polymers* 1984, **2**, 19.
[4] E. Högfeldt, *Ark. Kemi*, 1954, **7**, 561.
[5] E. Ekedahl, E. Högfeldt and L. G. Sillén, *Acta Chem. Scand.*, 1950, **4**, 556.
[6] W. J. Argersinger, A. W. Davidson and O. D. Bonner, *Trans. Kansas Acad. Sci.*, 1950, **53**, 404.
[7] E. Högfeldt, *Ark. Kemi*, 1952, **5**, 147.
[8] A. V. Mikulich, *Thesis*, Institute of General and Inorganic Chemistry, Akademia Nauk BSSR, Minsk, 1982.
[9] O. D. Bonner and V. Rhett, *J. Phys. Chem.*, 1954, **57**, 254.
[10] V. S. Soldatov and A. V. Mikulich, *Dokl. Akad. Nauk BSSR*, 1981, **35**, 724.

11

Industrial applications of solvent extraction

Michael Cox
Division of Chemistry, School of Natural Sciences, The Hatfield Polytechnic, Hatfield, Hartfordshire

Industrial applications of solvent extraction are extensive and range from the separation of aromatic from aliphatic components of petroleum feed stocks to the separation and recovery of actinides and uranium in nuclear fuel reprocessing. This article will concentrate on non-ferrous metal extraction and recovery. However other inorganic applications of this technique have been developed, for example the recovery of acids from spent pickling baths [1], and the purification of wet-process phosphoric acid for the food industry [2]. Considerable interest has recently been shown in the use of phosphoric acid as a feed stock for uranium [3] and other elements and a number of processes have been developed [4]. These aspects of the subject will not be further considered here.

A survey of the literature [5–7] reveals potential industrial processes for a wide range of elements, but the major contributions are concentrated on just a few metals.

11.1 COPPER

The development of commercial recovery processes for copper ensured the recognition of solvent extraction as an established hydrometallurgical unit operation in the non-ferrous metal industry and paved the way for its incorporation in flow-sheets for other metals. There are now many plants in operation throughout the world that use leach liquors from oxidic feed stocks [8] producing about 700 tonnes of copper per day (see Table 11.1). The extractants range from the hydroxyoximes, to carboxylic (Versatic) acid, and β-diketone (LIX54) for the ammoniacal leach liquors. This latter reagent is a weaker extractant than the hydroxyoximes, but it has high loading characteristics and is more easily stripped. It also has the advantage for this application that it does not extract ammonia. Commercial processes are also available for the recovery of copper from waste streams. Thus the versatility of solvent extraction to operate efficiently across a wide range of equipment sizes and process flow-rates has been used in the MECER process to recover copper from

Table 11.1 — Summary of copper solvent-extraction plants (1984).

Plant	Feed	Copper production (kg/day)	Comments
Ranchers Exploration and Development Co. Ltd., Miami, Arizona.	Dilute H_2SO_4, oxide ore	18200	World's first Cu/SX plant Uses LIX64N
Cyprus Bagdad Copper Co., Bagdad, Arizona.	Dilute H_2SO_4, oxide ore	18200	Uses LIX64N
Cyprus Mines Corp., Johnson Camp, Arizona.	Dilute H_2SO_4, oxide ore	13650	
Cities Services Co., Miami, Arizona.	Leach liquor from caved area of old underground mine	13650	Uses LIX64N in Holmes and Narver low profile M/S
Cities Services Co.	Dump leaching of oxide	15000	On stream mid 1981, uses LIX622(?)
Anamax, Twin Buttes, Arizona	Dilute H_2SO_4, oxide ore	90000	Twin stream plant, one uses PT5050, other LIX6022
Noranda Lakeshore, Arizona.	Dilute H_2SO_4, oxide ore	70000	On-stream 1981, uses P5100.
Duval Corp., Battle Mountain.	Heap leach of oxide	20000	Uses LIX64N
Inspiration Consolidation Copper Co., Arizona	Ferric cure leach of oxide	70000	Operational late 1979, uses P5100
Kennecott, Ray Mines Division, Arizona	Leach of silicate ore	100000	uses P5100
Mineroperu, Cerro Verde, Peru.	Oxide leach	90000	Uses LIX64N
Centromin, Cerro de Pasco, Peru.	Mine waters and dump leach	16000	Uses P5100
Cia. Minera de Cananea SA, Mexico.	Oxide ore dump leaching	40000	Uses P5100, on-stream 1980
Lo Aguirre Copper Mine, Socidad Minera Pudahuel Ltda, Chile.	Acid leach of oxide ore	50000	Operational date 1980
Matthey Rustenberg Refiners, Rustenberg, S. Africa.	Copper and nickel sulphate solutions from Merensky reef tailings	unknown	Uses Versatic acid for copper
Nchanga Consolidated Copper Mines, Nchanga, Zambia.	Acid leach of tailings	182000	Uses SME529 on one stream, LIX64N on remaining three.
Mitsui Corp., Takehara, Japan.	Ammoniacal leach of ISF copper dross	7600	Uses LIX64N
Sulphide Corp., Cockle Creek, Australia	Ammoniacal leach of ISF copper dross.	14000*	Uses LIX54, product copper sulphate crystals.
Commonwealth Smelting Ltd., Avonmouth, UK	Ammoniacal leach of ISF copper dross.	6000	Uses LIX54, operational 1981.
Johnson Matthey & Co. Ltd., London.	Copper and nickel sulphate solutions	unknown	Uses LIX64N

† kg/day of $CuSO_4 5H_2O$

spent ammoniacal etchants used in printed circuit board manufacture. The flow-sheet is shown in Fig. 11.1, and employs two stages of extraction. In the first stage,

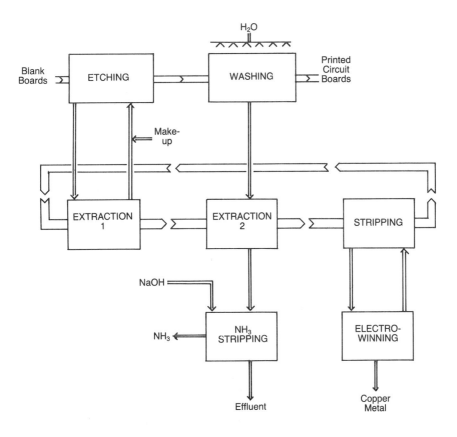

Fig. 11.1 — Flow sheet of the MECER process.

the copper concentration of the etchant is reduced from about 160 g/l to 100 g/l, the optimum level for etching. The loaded organic phase is then contacted with the rinse water removing the copper in this solution until the concentration is down to environmentally acceptable levels. The loaded copper is finally stripped with a recycle stream from the electrowinning cells. Plants are now operating in Germany, Sweden, U.K., and the USSR.

The advantage of the β-diketone extractant in this application was demonstrated in the development of the Imperial Smelting Furnace (ISF) process for the recovery of copper from copper–lead dross [9]. The initial test work used LIX64N, but replacement with LIX54 allowed a ten-fold reduction of plant size in addition to the benefits of eliminating ammonia extraction.

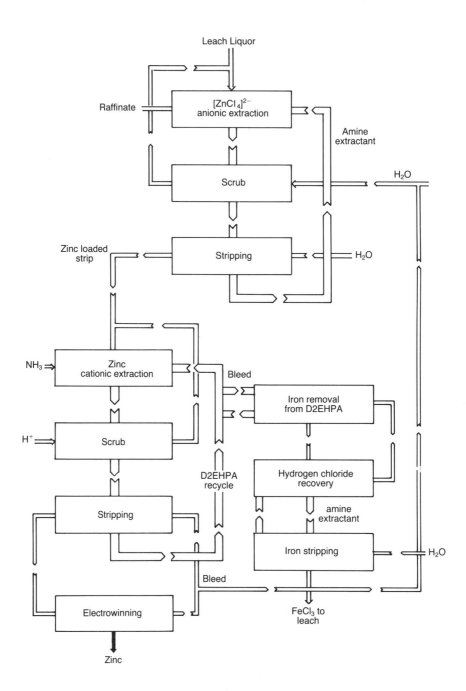

Fig. 11.2 — ZINCEX process flow-sheet (simplified).

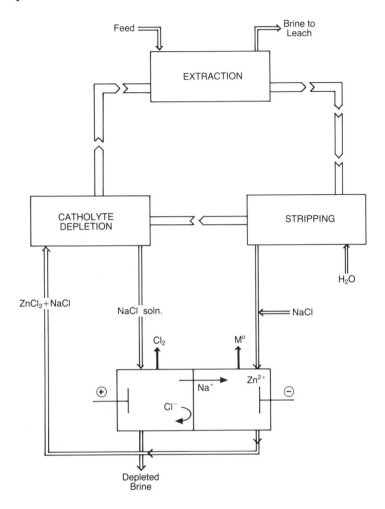

Fig. 11.3 — ZINCLOR process flow-sheet (simplified).

11.2 ZINC

The Zincex process for the recovery of zinc from pyrite cinders has been well documented [10]. The use of a two-stage extraction involving amine extraction of the chloride leach liquor followed by D2EHPA to purity the zinc from any iron co-extracted in the first stage means that care must be taken to ensure minimum entrainment of the amine into the second stage as this would tie up the extractants as amine salts (Fig. 11.2). A more recent development from Tecnicas Reunidas is the Zinclor process (Fig. 11.3) [11] where the zinc feed is leached with brine and then contacted with a new commercial solvating reagent di-pentylpentylphosphonate (DPPP) which extracts zinc as ZnL_2Cl_2. This reagent has high selectivity for zinc over other metals thereby giving a pure aqueous feed for electrowinning on stripping with hot water. Another interesting feature of this flowsheet concerns the use of a diaphragm cell for electrowinning. The aqueous strip solution is fed to the cathode compartment where the zinc is deposited on aluminium or titanium electrodes. Charge balance is achieved by transfer of sodium ions from the anode compartment through an ion-exchange membrane. Residual zinc chloride from the spent electrolyte is removed by solvent extraction with the barren organic phase to give a brine solution which is then recycled to the anode compartment before removal from the circuit. This process is capable of being used for other elements, e.g. copper.

11.3 COBALT AND NICKEL

The separation of these metals by using organophosphorus acids has been discussed earlier (Chapter 8), so only those processes involving chloride media will be outlined here. The main problems occurring with industrial processes is the variation of feed-stock composition, and often the flow sheets illustrate different ways of removing impurities. Consideration of this really takes the subject away from solvent extraction and into hydrometallurgy, but as an illustration the two approaches by Falconbridge Nikkelwerk and Metallurgie Hoboken will be outlined [12]. Both processes use a chloride leach followed by oxidation of iron and copper. The two flow sheets then diverge (Fig. 11.4). Falconbridge uses solvent extraction with tri-n-butyl phosphate (TBP) to remove iron from the leach liquor which has the approximate composition 120 g/l of nickel, 2 g/l of copper, cobalt, and iron, and 165 g/l of hydrochloric acid. Then the cobalt and copper are both removed by extraction with 10% tri-iso-octylamine leaving nickel in the raffinate. Selective stripping with water is then employed to separate cobalt and copper at a phase ratio O/A 30/1 and 20/1 respectively. The aqueous products are then further treated in the cobalt and copper refineries while the nickel is converted into nickel powder. The Metallurgie Hoboken circuit on the other hand uses precipitation and cementation processes to remove a number of contaminants. Thus iron is removed by alkali and copper by cementation with cobalt powder. Other contaminants can be removed similarly, leaving zinc, cobalt and nickel for solvent extraction treatment. Here, the ease of formation of chloro-complexes allows the separation of zinc at low chloride concentrations by amines; then after the chloride concentration is increased, cobalt can be separated from nickel.

11.4 PRECIOUS METALS

The classical approach to precious metal refining was based on chemical separations using traditional methods of precipitation and crystallization. However, although the process has been successfully operated for many years, the industry has been looking for alternatives because of the poor selectivity of some of the separations, and the many recycle streams necessary to obtain acceptable product purity. This involved a significant inventory of precious metals in process intermediates, which is an important factor in the overall economics of the process.

The chemistry of this group of metals is quite complex, with a number of oxidation states readily attainable under normal process conditions and, compared with the first row transition metals, relatively slow kinetics. Because of this, and the low profile and conservative attitudes of the industry, development of alternative processes has taken many years to reach commercial application. The three pub-

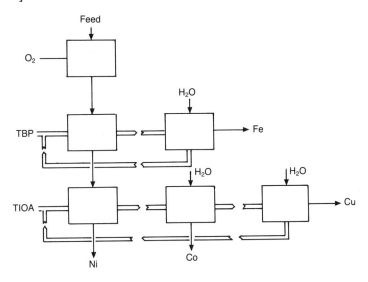

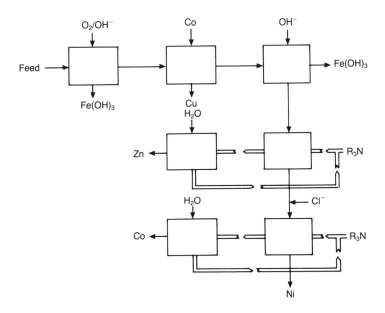

Fig. 11.4 — Simplified flow-sheets for cobalt/nickel separation from chloride media.

lished flow sheets differ quite considerably in both the order in which the metals are extracted and in the choice of extractant. As with other flow sheets, the route chosen will depend on the exact nature of the feed stock and, in particular, on the concentration of base metal contaminants and the relative abundance of individual precious metals.

In spite of the variables, there are a number of common features in these flow sheets (Fig. 11.5). All use a chloride leach that leaves silver in the residue, and all use

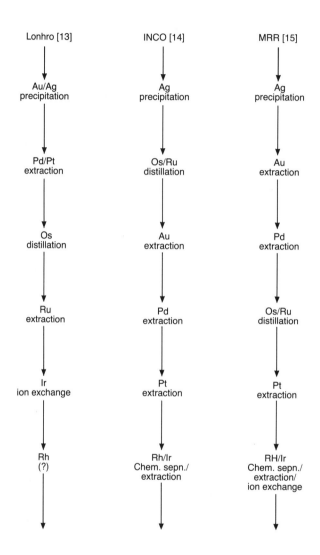

Fig. 11.5 — Simplified flow-sheets for precious metal processes.

the easiest way of removing osmium, distillation of the tetroxide, OsO_4, sometimes in combination with ruthenium, but the position of this operation in the overall flow sheet differs.

The major interest in these processes is, however, the choice of solvent extraction reagents and the way they are employed, and this will now be considered.

11.4.1 Gold

Dibutylcarbitol (2,2′-dibutoxydiethyl ether) for batch extraction of gold was first used commercially by INCO in 1971 [16]. Stripping of the loaded solvent was carried out by reduction with aqueous oxalic acid to produce gold grain directly. This system is retained in their current flow sheet. By contrast, MRR use methyl isobutyl ketone as the gold extractant. However, here contamination of the gold extract by iron and tellurium is greater than with dibutylcarbitol, but this does have some advantage in that these elements are removed early in the process. Gold is again obtained by reductive stripping, this time with iron powder.

11.4.2 Palladium and platinum

The Lonhro process is the only one to remove these elements together, with an amino-acid, R_2NCH_2COOH. This novel reagent has the advantage of good extraction and stripping characteristics for both elements, and avoided the problem of amines causing palladium retention in the organic phase. An interesting feature of the equipment was the use of a pulsed packed column for both the extraction and stripping operations. The exact process for palladium/platinum separation has not been specified, but mention was made of both sulphides, which were successfully used on a pilot scale, and oximes. Both the classes of reagent are used in the alternative flow sheets, thus INCO use di-n-octylsulphide and MRR uses a mixture of hydroxyoxime, again unspecified, with an amine. One of the problems with platinum-group metals is their relatively slow kinetics for ligand-transfer reactions; for this reason the INCO process uses a batch extraction for palladium recovery. A similar problem occurs with the oxime extractant, but here it has been found that addition of an amine acts as a kinetic accelerator and normal continuous operation is possible. The stripping of the loaded sulphide is carried out with ammonia, and the palladium is recovered as the chloroammine $[Pd(NH_3)_2Cl_2]$, via the $[Pd(NH_3)_4]Cl_2$ species. In the other process the palladium is stripped by hydrochloric acid.

$$\overline{Pd(R_2S)_2Cl_2} + NH_3 \rightarrow [Pd(NH_3)_4]^{2+} + \overline{2R_2S} + 2Cl^-$$

The next element to be extracted is platinum, and again different approaches have been taken by INCO and MRR. INCO use TBP to extract platinum from 5–6M hydrochloric acid, the loaded phase is scrubbed with more hydrochloric acid, then finally stripped by water to give chloroplatinic acid, H_2PtCl_6. The alternative is to use an amine to extract $[PtCl_6]^{2-}$ as an ion pair (MRR).

$$\overline{2R_3N.HCl} + [PtCl_6]^{2-} \rightleftharpoons \overline{[R_3NH]_2[PtCl_6]} + 2Cl^-$$

This does present some difficulties in that the equilibrium shown lies well to the right, so that strong acid or alkali is required for stripping. Another disadvantage is the possibility of amine replacing chloride in the complex species to form inner compounds such as $[Pt(NH_3)_nCl_{6-n}]^{(2-n)-}$, which would lead to a build up of platinum in the organic phase. In both the processes prior reduction to reduce any iridium to iridium(III) is required to prevent it from being co-extracted with the platinum.

The separation of the remaining platinum-group metals still poses some difficulties. Iridium can be extracted after oxidation to iridium(IV) by either TBP (Lonhro, INCO) or amines (MRR) and, in the latter case, reduction to iridium(III) (which does not form stable ion-pairs) can provide an easy means of stripping [17]. The Lonhro process, which does not remove ruthenium by distillation, instead extracts it as a nitrosylruthenium complex with a tertiary amine, followed by stripping with alkali [18]. Finally, rhodium is removed by ion exchange (Lonhro, MRR) to leave a raffinate consisting of base metals. Because of the intrinsic value of precious metals, care is taken to ensure very low concentrations in raffinate solutions. Rigorous disentrainment of organic compounds from these raffinates has also proved to be very important to prevent contamination of subsequent process streams. In this industry, because of the complexity of the aqueous-phase chemistry, rigorous control is necessary throughout the process of variables such as acidity, redox potential, metal-ion concentration etc., but because of the experience of the companies and operators in this field and the relatively small size of operation this has not proved to be too difficult.

11.5 CONCLUDING REMARKS

Solvent extraction is now a well-established operation in hydrometallurgy for a wide variety of metals. The ability to deal with a variety of feed stocks and solutions is an advantage, as is the ability to operate successfully over a wide range of scale and flow-rate.

What of the future? One of the exciting new developments concerns liquid membranes. Of the two alternatives, it is likely that the polymer-supported membranes will be in commercial operation before the liquid-surfactant membranes, because the technology is very similar to that of the established techniques of reverse osmosis and ultrafiltration. The reduction in volume of the organic phase in this system has obvious economic advantages, but a number of problems still remain to be solved. However, the potential to operate with unclarified liquors and perhaps solvent-in-pulp will provide the incentive for development. Therefore, with solvent extraction now playing a vital role in hydrometallurgy, if the demand for metals increases an expansion in activity in this area can be expected both in construction of plant and commercial application of processes now at bench/design stages.

REFERENCES

[1] T. Lo, M. Baird and C. Hanson (eds.), *Handbook of Solvent Extraction*, Wiley, New York, 1983, Chapter 25.10.
[2] Ref. [1], Chapter 26.
[3] Ref. [1], Chapter 25.11.
[4] *Proceedings, 2nd International Congress on Phosphorus Compounds, Boston, Mass.*, Institut Mondail du Phosphate 1980, Theme IV.
[5] D. S. Flett, *Chem. Eng.* 1981, **370**, 321.
[6] G. M. Ritchey and A. W. Ashbrook, *Solvent Extraction, Part II*, Elsevier, Amsterdam, 1979.
[7] Ref. [1], Chapters 24, 25.1–25.14, 26.
[8] D. S. Flett, in *Hydrometallurgy, Research, Development and Plant Practice*, Osseo-Asare and Miller, eds., AIME, 1984, 357.
[9] W. Hopkins, Paper given at 105th AIME meeting, Las Vegas, 1976.
[10] E. D. Nogueira, J. M. Regife and P. M. Blythe, *Chem. and Ind.* **1980**, 63.
[11] E. D. Nogueira, L. A. Suarez-Infanzon and P. Cosmen, paper presented at 'Zinc 83', 13th Annual Hydrometallurgical meeting, CIM, Edmonton, Canada, 1983.
[12] Ref. [1], Chapter 25.2.

[13] R. I. Edwards, *ISEC77, International Solvent Extraction Conference, Proceedings*, Toronto, Canada, Vol. 1, 1979, p. 24.

[14] J. E. Barnes and J. D. Edwards, *Chem. and Ind.*, **1982**, 151.

[15] L. R. P. Reavell and P. Charlesworth, *ISEC80, International Solvent Extraction Conference, Proceedings*, Liege, 1980.

[16] B. F. Rimmer, *Chem. and Ind.*, **1974**, 63.

[17] M. J. Cleare, P. Charlesworth and D. J. Bryson, *J. Appl. Chem. Biotech.*, 1979, **29**, 210.

[18] Anon., *Chem., Eng., 1979,* **83**, 90.

12

Solvent extraction in the nuclear industry†

Pier R. Danesi*
Chemistry Division, Argonne National Laboratory, 9700 South Cass Avenue, Argonne, Illinois 60439

12.1 INTRODUCTION

The industrial use of solvent extraction was pioneered by the nuclear industry in the early forties. The high solubility of uranyl nitrate in diethyl ether, a property that had been known to chemists for many years, was exploited by the nuclear scientists for the large scale purification of uranium, required in a super-pure state for nuclear reactions. The same property was exploited for the second purification of plutoniun, performed in Chicago in 1942. This was the beginning of a large number of applications of solvent extraction to the separation and purification of metals both in the nuclear industry and in extractive metallurgy.

Solvent extraction is probably one of the most powerful purification technologies and it is particularly suitable for nuclear applications, where very high recoveries and separation factors are required. More specifically solvent extraction is very suitable for nuclear processes because:

(a) it is a multi-stage operation
(b) it can be easily engineered for remote control
(c) it generates small volumes of wastes
(d) the radiolytic degradation of the solvent can be easily compensated in a continuous way.

Today the most used solvent-extraction reagent of the nuclear industry is no longer diethyl ether, which presents many problems because of its volatility and flammability. Since 1950, the leading solvent of the nuclear industry has been tri-n-butylphosphate, TBP, which is used in several different nuclear processes. The popularity of TBP is due to its continuous and successful use in the most important

† Work performed under the auspices of the Office of Basic Energy Sciences, Division of Chemical Sciences, U.S. Department of Energy under contract number W-31-109-ENG-38.
* Present address: International Atomic Energy Agency, Seibersdorf Laboratory, Vienna, Austria.

and challenging solvent extraction process, i.e. the PUREX process. The name PUREX stands for Plutonium Uranium Recovery by EXtraction. This process has been in use for more than 30 years in several countries to recover and purify plutonium and uranium from dissolved irradiated nuclear fuels both in civilian and military reprocessing plants.

In this article, only a few of the very many solvent extraction processes of the nuclear industry will be briefly illustrated. The processes have been chosen with the only objective to underline to the reader the great importance that solvent extraction plays today in the various phases of the nuclear industry. The examples of solvent-extraction processes selected deal with:

(a) the extraction and purification of uranium from its ores,
(b) the reprocessing of irradiated nuclear fuel to recover and purify uranium and plutonium from fission products,
(c) the removal of transuranium elements and fission-product rare earths from high-level liquid wastes,

No attempt has been made to give a comprehensive treatment of a subject which would require a series of books.

Most of the emphasis has been put on the PUREX process, which is one of the most important solvent extraction processes used in the technological world.

12.2 IRRADIATED NUCLEAR FUEL [1,2]

The solvent extraction of uranium and plutonium by TBP, and their separation from the fission products and each other, is just one of the many steps which are contained in the fuel cycle. Figure 12.1 gives a schematic description of the entire fuel cycle. The cycle involves the handling of the fissile and fertile materials which are necessary for the production of nuclear energy, as well as the radioactive products generated during the operation of the nuclear reactor. The fuel cycle is conventionally divided into a front- and a back-end. The borderline between the two parts crosses the nuclear power station. The front-end involves uranium exploration, mining and refining, isotope enrichment and fuel-element fabrication. Solvent extraction is used in the front-end for the extraction and concentration of uranium from its ores and two processes having this objective will be described later in this paper.

The back-end involves the reprocessing of the irradiated nuclear fuel and the treatment of the radioactive wastes. In both these steps, solvent extraction plays a very important role.

Fuel elements containing uranium-238 at different degrees of enrichment in ^{235}U are 'burned' in the nuclear reactor. During the functioning of the nuclear reactor, ^{235}U is partly transformed into fission products as a result of the fission reaction. At the same time various actinide isotopes are generated from ^{235}U and ^{238}U through a number of nuclear reactions. Radioactive decays further increase the complexity of the mixture of chemical elements and isotopes.

The chemical composition of the spent nuclear fuel varies with the nature and amounts of fissile and fertile materials, the neutron energy and flux, the exposure time, and the cooling time.

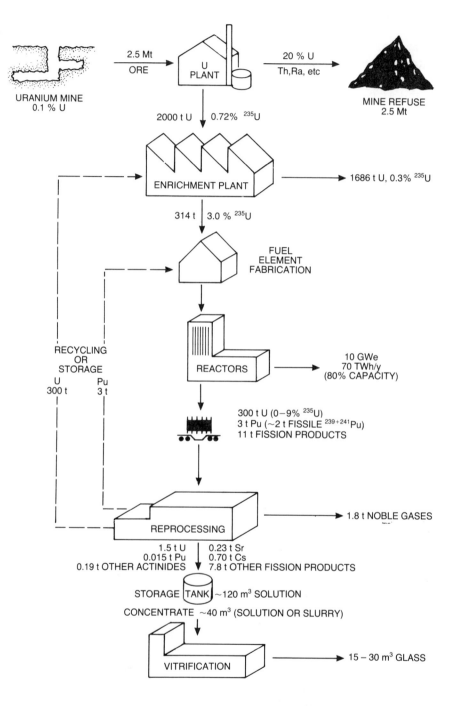

Fig. 12.1 — Annual flow of materials in a 10 GWe LWR fuel cycle program. Reprinted with permission from G. R. Choppin and J. Rydberg, *Nuclear Chemistry*, © 1980, Pergamon Press Ltd.

Different chemical and isotopical compositions can therefore be expected when fuel elements from different reactors (PWR, BWR, HWR, GCR, etc.), with different irradiation and cooling times, are treated. However the differences in chemical composition are never too large, and the PUREX process is capable of reprocessing all fuel elements with only minor modifications. As an example Table 12.1 reports the total amount of plutonium formed in different types of reactors. It is

Table 12.1 — Production of Pu (kg/MWe y) in various nuclear reactors

Nuclear reactor	Total Pu	Fissile Pu
Light water	0.26	0.18
Heavy water	0.51	0.25
Gas-cooled	0.58	0.43
Fast breeder	1.35	0.7 to 1.0

clear from the table that heavy-water and gas-cooled reactors are the best producers of plutonium. For this reason they are used to obtain the Pu needed for weapon fabrication.

The radioactivities and the amounts of the various fission products and actinides present in 1 tonne of spent uranium fuel are shown in Fig. 12.2. The degree of enrichment burn up, neutron flux and cooling time of the fuel are reported in Table 12.2. The quantities and radioactivities of the main actinides are reported in Table 12.3. From Fig. 12.2 it appears that if the corrosion products Cr, Fe, Ni are included, 47 different chemical elements, out of a total of 103, can be found in a solution obtained by dissolving the spent nuclear fuel. In the reprocessing of the nuclear fuel, just two chemical elements, Pu and U, have to be separated out of this complex mixture.

The fission products elements which are formed in the largest amounts (~70%) are Xe, Zr, Mo, Nb, Cs and Ru (the relative amounts of the various fission products and the relative radioactivities of the various fission products, after a given irradiation time, at different cooling times, can be calculated). After a cooling rime of 10 years the major contribution to the activity of the fission products derives from ^{90}Sr and ^{137}Cs. For cooling times longer than 10^3 years the activity of the fission products becomes practially independent of the cooling time. The activities of the most important fission products after a cooling time of 10^3 years are reported in Table 12.4.

In practice, the cooling time of the nuclear fuel, before reprocessing is started, is a compromise between (a) the interest loss in the immobilized capital represented by the valuable fissile and fertile materials and storage cost, on one side, and (b) the economical incentive to reprocess less radioactive fuels on the other. Presently military fuel is reprocessed after six months while fuel from commercial power reactors is reprocessed after about ten years. This long period of time is due to limitations of the existing reprocessing capabilities.

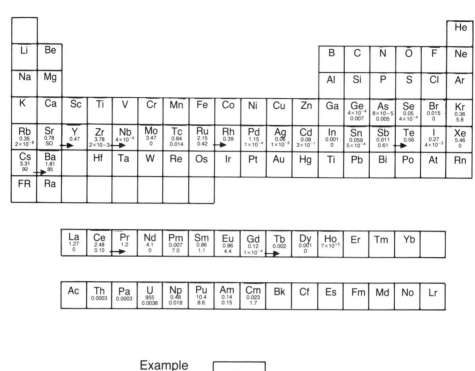

Fig. 12.2 — Composition of spent uranium fuel. Reprinted with permission from G. R. Choppin and J. Rydberg, *Nuclear Chemistry*, © 1980, Pergamon Press Ltd.

Table 12.2 — Properties of spent uranium fuel

1 tonne spent U fuel
3.3% enriched in ^{235}U
Burnup: 33 000 MW day/tonne of uranium
Flux: 3×10^{13} neutrons cm^{-3} sec^{-1}
Cooling time: 10 years

Total radioactivity
F.P.:   312 k Ci
T.U.:   2.35 k Ci

Thermal power
F.P.:   1.02 kW
T.U.:   0.070 kW

Table 12.3 — Amounts and radioactivities of the main actinides formed in 1 tonne of spent uranium fuel after a cooling time of 10 years

Actinide	Weight (kg)	Radioactivity (k Ci)
Th	—	3×10^{-4}
Pa	—	3×10^{-4}
U	955	3.6×10^{-3}
Np	0.48	1.8×10^{-2}
Pu	10.4	8.6
Am	0.14	0.16
Cm	0.023	1.7

Table 12.4 — Activities of long-lived fission products after a cooling time of 10^3 years

Isotope	Half-life (years)	Radioactivity (Ci per tonne of U)
^{79}Se	6.5×10^4	0.39
^{87}Rb	4.7×10^{10}	2×10^{-5}
^{93}Zr	1.5×10^{10}	2.0
^{99}Tc	2.1×10^5	14.3
^{107}Pd	6.5×10^6	0.11
^{126}Sn	10^5	0.54
^{129}I	1.6×10^7	0.038
^{135}Cs	2.1×10^6	0.29

12.3   REPROCESSING OF URANIUM FUEL ELEMENTS FROM THERMAL REACTORS [3—5]

The first reprocessing of nuclear fuels was dictated by military desire to fabricate nuclear weapons. Later, incentives to reprocess nuclear fuel also came from the need to conserve finite energy resources and to make easier and safer the management of the nuclear wastes.

The nuclear industry is presently mainly based on the thermal fission of ^{235}U. Uranium-235 is present in natural uranium (major isotope ^{238}U) to the extent of 0.72%. Uranium-235 is split in a fission reactor by a neutron to generate energy, fission products and more neutrons. These can in turn cause more fissions. The surplus neutrons are absorbed by structural materials, by shields and by ^{238}U, which is in turn transformed into ^{239}Pu. Plutonium-239 can itself undergo fission. The fuel element has to be removed from the nuclear reactor when only a fraction of the total ^{235}U has been used up because of the adverse effect produced by the accumulation of the neutron absorbing fission products and the structural deterioration of the fuel elements. The objective of fuel reprocessing is therefore the recovery of the remaining uranium (^{238}U + ^{235}U), the recovery of the ^{239}Pu produced, which can be used as a substitute fissionable material, and their purification from the fission products.

Before the chemical separation is started the nuclear fuel is chopped up by suitable cutting machines and then dissolved in boiling $6–11\,M$ HNO_3, in stainless steel vessels. To improve the dissolution, fluoride ions are added to the nitric acid at a concentration of $\leqslant 0.05\,M$. The main steps of the reprocessing of the nuclear fuel are shown schematically in Fig. 12.3. The incentive to reprocess nuclear fuels derives

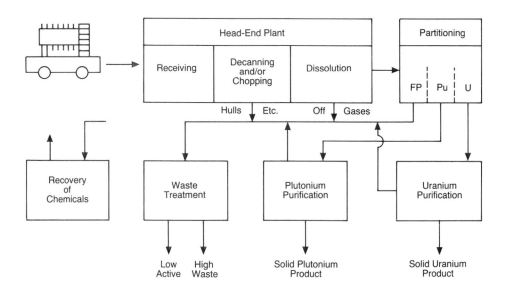

Fig. 12.3 — Main steps in reprocessing spent LWR fuels. Reprinted with permission from G. R. Choppin and J. Rydberg, *Nuclear Chemistry*, © 1980, Pergamon Press Ltd.

from requirements to increase the energy extracted from fissile and fertile material, reduce the hazards and cost of handling the nuclear wastes, reduce the cost of the fuel cycle, and extract valuable products (^{90}Sr, ^{137}Cs, etc.) from the active wastes..

The reprocessing of the nuclear fuel is one of the most challenging activities of the chemical industry, since very stringent requirements are imposed on it. These requirements are the following:

(a) a high recovery (99% or more) of U and Pu must be obtained because of the high value of these materials and in order to simplify the waste management,
(b) the process must be flexible enough to deal with the complexity of the mixture of chemical elements,
(c) high decontamination factors of U and Pu from fission products (10^6 to 10^8) must be obtained

(d) high technical reliability is necessary because of the difficulty and high cost of repairs,

(3) it is necessary to operate in a remote way,

(f) criticality must be avoided,

(g) no waste streams can be discharged.

All the above requirements are met by the currently operated solvent extraction PUREX processes. In view of this exceptional performance, in spite of these stringent requirements, the PUREX process probably deserves to be described as the 'most important solvent-extraction process of the nuclear industry'.

12.4   CHEMISTRY OF THE PUREX PROCESS [3–6]

12.4.1   Tri-n-butyl phosphate (TBP)

TBP in the pure state is an odourless and colourless liquid. The commercial product sometimes has a slight yellow-brown cast. TBP is used in PUREX-type processes diluted with inert diluents that reduce the density, viscosity and the extractive power to suitable levels. The most commonly used diluents are refined kerosene (aliphatic) or mixtures of n-paraffins. The proportions of TBP in the diluent may vary from 5 to 30% by volume according to the different versions of the process. The properties of undiluted TBP are shown in Table 12.5. When TBP extracts uranyl nitrate, two

Table 12.5 — TBP, $(CH_3-CH_2-CH_2-CH_2-O)_3P=O$

Density	0.973 g/cm^3
n_D	1.4223
Viscosity (cP)	33.2
Solubility in water	0.39 g/l.
Water solubility in TBP	64 g/l.
Dielectric constant	7.97
Flash point	294°F

water molecules are displaced per mole of extracted uranyl nitrate. When nitric acid is extracted the water content of the organic phase goes through a maximum when the concentration of the aqueous HNO_3 is between 2 and 5 M.

TBP acts as an extractant because of the basic properties of its phosphoryl group P=O. The alkoxy oxygens do not play any direct role in the extraction. When metals are extracted by TBP a covalent bond is formed. Acid and water are bonded to the phosphoryl group of TBP by hydrogen bonds.

TBP is a very stable organic compound, but nevertheless it can hydrolyse when in contact with aqueous solutions. During the hydrolysis, small amounts of di-n-butylphosphoric acid, DBP (i.e. $(RO)_2 PO-OH$ with R=butyl) are formed. The formation of DBP is the first step of the dealkylation process which eventually leads to the formation of the completely water-soluble orthophosphoric acid. The formation of DBP has a very deleterious effect on the behaviour of TBP in the PUREX process. DBP increases the extraction coefficient of U and Pu at low acidity to the point that their stripping from the organic phase becomes extremely difficult. Zirconium is also very strongly extracted by DBP, and the decontamination of the organic diluent from this element becomes a serious problem when DBP is present. DBP is also formed as a result of the radiolysis of TBP. In addition other polymeric compounds are formed between DBP and Pu(IV) and Zr(IV). The amount of DBP formed by radiolytic degradation is approximately proportional to the amount of energy absorbed. For this reason there is now a trend towards the use of solvent extraction contactors which minimize the contact time of the aqueous and organic phases (centrifugal contactors). To eliminate DBP and the other degradation products from TBP, the PUREX solvent is continuously purified by a sequence of basic (sodium hydroxide or sodium carbonate), water, and acidic washes. In this way, DBP and traces of extracted fission products are removed and the solvent can be recycled for further use. In practice the TBP lost through degradation, solubility and aqueous entrainment is replaced by addition of fresh extractant.

TBP is an excellent all-round solvent for the reprocessing of irradiated nuclear fuel. It possesses all the major requirements which a solvent must have to be successfully used in an industrial solvent-extraction separative process. The properties which make TBP such a good solvent for nuclear fuel reprocessing are listed below:

(a) it extracts selectively the valuable elements (U and Pu) leaving in the aqueous phase the fission products and inert elements (Fe, Cr, Ni),
(b) the distribution ratios of the extracted species and the solvent loading are high enough to allow the U and Pu to be completely recovered at moderate phase ratios,
(c) the extractant is adequately stable against chemical and radiolytical degradation,
(d) the density, viscosity and interfacial tension of 25–30% TBP in a number of diluents permit rapid mixing and separation of the phases,
(e) the volatility and water-solubility are sufficiently low to prevent significant losses,
(f) the flash point is high enough to reduce fire hazards to a minimum,
(g) it is not toxic,
(h) it is easily synthesized, readily available and inexpensive.

12.4.2 Extraction of nitric acid by TBP
Nitric acid is extracted by TBP according to the reaction

$$H^+ + NO_3^- + \overline{TBP} \rightleftharpoons \overline{TBPHNO_3}$$

where the bar indicates organic-phase species. When the concentration of the aqueous HNO_3 exceeds $7M$ the HNO_3/TBP mole ratio in the organic phase becomes larger than one, because of formation of $TBP(HNO_3)_n$ species. Uranyl nitrate and other extractable metal nitrates reduce the extraction of HNO_3 by TBP since they simultaneously compete for the same phosphoryl group.

12.4.3 Uranium

Uranium generally exists in aqueous solutions in the hexavalent state as uranyl ion, UO_2^{2+}, unless strong reducing agents are present. Uranyl nitrate is readily extracted by TBP according to the reaction

$$UO_2^{2+} + 2NO_3^- + \overline{2TBP} \rightleftharpoons \overline{(TBP)_2UO_2(NO_3)_2}$$

The equilibrium constant for the extraction reaction, K, and the distribution ratio, E, are given by the equations

$$K = \frac{[\overline{(TBP)_2UO_2(NO_3)_2}]}{[UO_2^{2+}][NO_3^-]^2[\overline{TBP}]^2}$$

$$E = \frac{[UO_2^{2+}]_{org}}{[UO_2^{2+}]_{aq}} = K[NO_3^-]^2[\overline{TBP}]^2$$

The extracted complex has a structure in which uranium has six co-ordination positions in its equatorial plane and the two uranyl oxygens at the poles. The $U-O$ bonds at the poles are shorter than the bonds between U and the equatorial oxygens. The bond angles between the equatorial oxygens are not rigidly fixed at $60°$ and can slightly vary. Each nitrate occupies two equatorial co-ordination sites. The remaining two sites are occupied by TBP. When uranyl nitrate is present in water solutions the two co-ordination positions occupied by TBP are occupied by two water molecules.

12.4.4 Plutonium

Plutonium chemistry is characterized by a variety of oxidation states which can coexist in aqueous solution. The redox potentials of plutonium in $1M\ HClO_4$ are the following:

$$\overset{\displaystyle 1.043}{\overline{Pu^{3+}\underset{0.982}{\rule{1.5cm}{0.4pt}}Pu^{4+}\underset{1.72}{\rule{1.5cm}{0.4pt}}PuO_2^+\underset{0.913}{\rule{1.5cm}{0.4pt}}PuO_2^{2+}}}$$

Although these potentials strictly apply only to $1M\ HClO_4$ they provide useful indications also for nitric acid medium. At the acidities which are met in the PUREX process plutonium, is completely converted into Pu(IV) by adding nitrous acid. The nitrite ion has the capability of oxidizing Pu(III) and simultaneously reducing Pu(VI). Pu(IV) can be reduced easily to Pu(III) by numerous reductants such as ferrous sulphamate, hydroxylamine or uranium(IV).

Pu(IV) hydrolyses extensively and polymerizes at low acidities. Since the hydrolysed and polymerized species are very poorly extracted by TBP and the hydrolysis and polymerization reactions are slow and difficult to reverse, low acidities have to be avoided in the processing of Pu(IV) solutions.

Pu(IV) is extracted by TBP according to the reaction

$$Pu^{4+} + 4NO_3^- + 2\overline{TBP} \rightleftharpoons \overline{(TBP)_2Pu(NO_3)_4}$$

The equilibrium constant for the extraction reaction, K, and the distribution ratio, E, are given by the equations

$$K = \frac{[\overline{(TBP)_2Pu(NO_3)_4}]}{[\overline{TBP}]^2[Pu^{4+}][NO_3^-]^4}$$

$$E = \frac{[Pu^{4+}]_{org}}{[Pu^{4+}]_{aq}} = K[\overline{TBP}]^2[NO_3^-]^4$$

The order of extractability of the various plutonium oxidation states by TBP is Pu(IV) > Pu(VI) ≫ Pu(III). PuO_2^{2+} is less extracted by TBP than UO_2^{2+} and NpO_2^{2+}.

12.4.5 Neptunium

Neptunium is characterized by the good stability of the pentavalent neptunyl ion, NpO_2^+. However tetravalent and hexavalent neptunium ions, Np^{4+} and NpO_2^{2+}, can be easily prepared and stabilized. The redox potentials of neptunium in $1M$ HClO$_4$ are

$$\overline{Np^{4+}\underset{0.739}{\overset{\overset{\displaystyle 0.938}{\rule{4cm}{0.4pt}}}{\rule{0pt}{0pt}}}NpO_2^+\underset{1.137}{}NpO_2^{2+}}$$

These can be used to make qualitative predictions about the valence state of neptunium in nitric acid environments. For example, when the HNO$_3$ concentration is between 3 and 4M more than 90% of Np is present as Np(V). Np(V) can be easily reduced to Np(IV).

The reoxidation of Np(IV) to Np(V) is a very slow process even when a holding reductant is not present. To oxidize Np(V) to Np(VI) strong oxidants are required.

Np(V) is very little extracted by TBP and in most PUREX process schemes neptunium is rejected to the aqueous waste as NpO_2^+. Neptunium is kept in the pentavalent state by the same nitrite ion which is responsible for the stabilization of tetravalent plutonium. In some versions of the PUREX process Np is extracted into TBP as $\overline{(TBP)_2Np(NO_3)_4}$. This result is obtained by adding to the feed solution a suitable reductant such as ferrous sulphamate. In this case plutonium is rejected to the waste as the nonextractable Pu(III).

12.4.6 Ruthenium

Ruthenium is, together with zirconium, one of the major problem elements of the PUREX process with reference to decontamination of the recovered U and Pu.

Ruthenium can exist in nitric acid in a multiplicity of species. Although the bulk of the Ru is easily separated from U and Pu, a small fraction of it is often transferred to the organic phase. The slow equilibria of Ru and its complicated chemistry are the origin of the difficulties. The problems arise from the formation of the nitrosylruthenium complexes formed during the dissolution of the nuclear fuel in nitric acid. The stable nitrosylruthenium species, $RuNO^{3+}$, has five positions which are available for co-ordination. Fortunately the dominant species are the poorly extractable (by TBP) complexes $[RuNO(H_2O)_5]^{3+}$ and $[RuNO(H_2O)_4(NO_3)]^{2+}$. Other $RuNO^{3+}$ species, such as $RuNO(H_2O)_2(NO_3)_3$ and $RuNO(H_2O)_3(NO_3)_2^+$, are more extractable. TBP extracts these two species, initially forming outer-sphere complexes. TBP slowly replaces the water molecules in the primary coordination sphere of Ru, and the Ru is slowly transformed into the very organophilic species $\overline{RuNO(TBP)_2(NO_3)_3}$ which is very difficult to remove from the organic phase. Since all the equilibria leading to the formation of the extractable Ru species are slow, rapid extractions, done in centrifugal contractors, together with slow scrubbing, improve the decontamination factors from Ru. The Ru isotopes which cause problems are ^{103}Ru and ^{105}Ru.

12.4.7 Zirconium

^{95}Zr is another radionuclide that causes decontamination problems in the PUREX process. Zirconium chemistry is characterized by the formation of several chemical species, with a slow equilibrium between them. In acid solutions the following soluble hydrolysed species exist:

$$Zr^{4+} \rightleftharpoons Zr(OH)^{3+} \rightleftharpoons Zr(OH)_2^{2+} \rightleftharpoons Zr(OH)_3^+ \rightleftharpoons Zr(OH)_4$$

These species can readily polymerize with time and on increasing the temperature. In the presence of nitrate ions, several complexes are slowly formed.

$$Zr(OH)_2NO_3^+ \rightleftharpoons Zr(OH)(NO_3)_2^+ \rightleftharpoons Zr(NO)_2^{2+}$$
$$\Updownarrow$$
$$Zr(OH)_2(NO_3)_2$$

Zirconium is strongly extracted by TBP to form the species $\overline{Zr(NO_3)_4(TBP)_2}$ and $\overline{Zr(NO_3)_4(TBP)_3}$. Fortunately the kinetics of all reactions is very slow, and by reducing the contact time between the organic and the aqueous phase, reasonable decontaminations of Pu and U from ^{95}Zr are obtained. ^{95}Nb behaves in a way similar to ^{95}Zr.

12.5 THE PUREX PROCESS [3—6]

The PUREX process is based on the ability of TBP to extract uraniun(VI), plutonium(IV and VI) and neptunium(IV and VI) from nitric acid solutions while leaving in the aqueous solution the fission products. Actinides in the tervalent state are only weakly extracted by TBP. Of all the fission products, only Zr (and Nb) and Ru are sufficiently extracted to create a problem. Tervalent rare earths are very little extracted. Technetium extracts only to a small extent as Tc(VII).

The distribution ratios of uranium(VI) and plutonium(IV) between 30% TBP in kerosene and aqueous HNO_3 are shown in Fig. 12.4. The distribution ratios of some

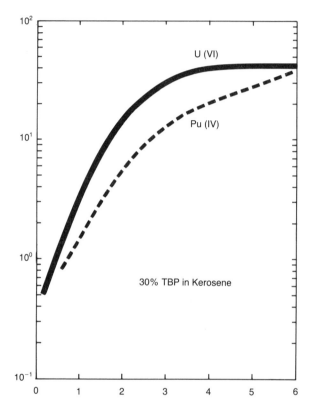

Fig. 12.4 — Distribution ratio of uranium(VI) and plutonium(IV) between 30% TBP in kerosene and aqueous nitric acid.

selected ions between HNO_3 and TBP are also reported in Table 12.6. Since Pu(III) is very little extracted and the distribution ratio of U(VI) dramatically decreases on lowering the aqueous HNO_3 concentration, a good separation of U and Pu from the various actinides and fission products is obtainable by using different acidities in the extraction and stripping steps and by using selective reductants.

The feed solution, obtained by dissolving the nuclear fuel in HNO_3, generally contains uranium at a concentration of about $1M$. Typical feed compositions from different reactor fuels are shown in Table 12.7.

Table 12.6 — Distribution coefficients of U, Pu and selected fission
products between $1 M$ HNO_3 and 30% TBP in kerosene at 25°C

Species	E
UO_2^{2+}	8.1
Pu^{4+}	1.55
PuO_2^{2+}	0.62
HNO_3	0.07
Zr	0.02
Ce^{3+}	0.01
Ru	0.01
Pu^{3+}	0.008
Nb	0.005
Rare earths	0.002
Combined beta emitters	< 0.0001

Once uranium and plutonium have been separated from the bulk of the fission products and the other actinides by extracting them into TBP, their separation is achieved by selectively stripping uranium and plutonium into aqueous solutions. The ability of Pu(IV) to be easily reduced to the nonextractable Pu(III) is the basis of the U–Pu separation.

In the initial feed solution uranium is always present as the stable hexavalent uranyl ion, UO_2^{2+}, whereas plutonium can be present in the 3, 4 and 6 oxidation states. Taking advantage of the fact that on addition of NO_2^- ions, Pu is stabilized in the extractable Pu^{4+} state, and UO_2^{2+} ions remain unaffected, both U and Pu are initially extracted into TBP. Plutonium is removed from the organic phase by reducing it to the nonextractable Pu^{3+} by contacting the organic phase with an aqueous solution containing ferrous ions or hydroxylamine. Once plutonium is removed, uranium is transferred back to the aqueous phase with a dilute HNO_3 solution. The low value of the U(VI) distribution ratio at low HNO_3 concentrations is exploited here. The principle of the separation is shown schematically in Fig. 12.5. The extraction and stripping procedures can be repeated to improve the separation; in this way it is possible to obtain very high decontaminations of U and Pu from fission products, very pure U and Pu products and losses of U and Pu of less than 1%. In general, the decontamination factors that are required from the gamma activity of the fission products are about 10^6 for uranium and 10^7 for plutonium. A very good separation of the two products is also required. The U:Pu ratio in the final product must be 10^8:1, if uranium has to be handled by using the same precautions as for natural uranium. The amount of uranium which can be tolerated in the separated plutonium is much higher. Generally the Pu:U ratio can be close to 10^3:1. In order to accomplish such separations, the scheme of Fig. 12.5. must be repeated at least twice, as indicated in Fig. 12.6, where the complete flowsheet of the PUREX process is shown. During the first extraction more than 99.9 mass % of U and Pu are extracted by TBP. Most of the contaminating fission products are only very slightly extracted. The few which are extracted are washed out of the organic phase by scrubbing it with a stream of HNO_3. Stripping of plutonium is generally achieved by use of a aqueous solution of ferrous sulphamate, $Fe(NH_2SO_3)_2$ or hydroxylamine. In

Table 12.7 — Feed compositions from various reactor fuels

Fuel type	Magnox	Thermal oxide	Fast reactor
Irradiation, MW day/tonne	3500	37 000	52 000 (mean)
Cooling time, months	6	12–60	6
Uranium molarity	1.25	1.25	1
Percentage of ^{235}U (approx.)	0.3	1–2	<0.2
Plutonium molarity (approx.)	0.002	0.01	0.16
Zirconium molarity $\times 10^4$	13	130	124
Ruthenium molarity $\times 10^4$	6.5	71	125
Caesium molarity $\times 10^4$	6.4	62	125
Strontium molarity $\times 10^4$	3.3	32	24
Iodine molarity $\times 10^4$	0.63	6.7	15
Total fission products Ci/litre	100	150–820	Ca. 3000
Neptunium molarity $\times 10^4$	0.22	5.8	1.5
Americium molarity $\times 10^4$	0.12	2.9	17
Curium molarity $\times 10^4$	0.0014	0.66	0.53

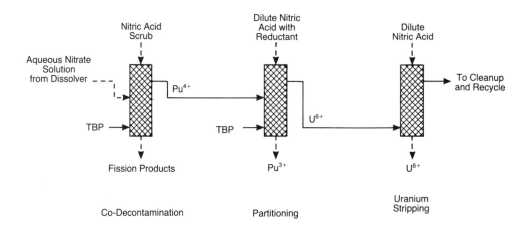

Fig. 12.5 — Schematic representation of extraction and stripping steps in a PUREX process. Dodecane containing 15–30 wt % of tributyl phosphate (TBP).

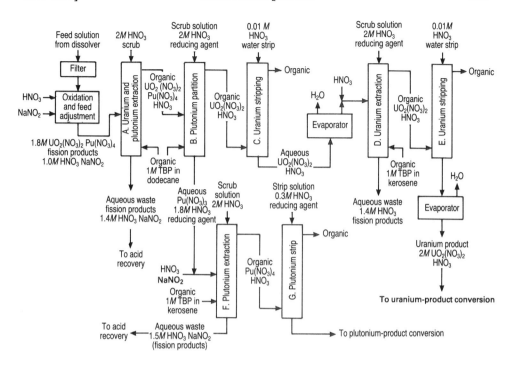

Fig. 12.6 — Flow sheet of PUREX process for separation of uranium, plutonium, and fission products by solvent extraction with tributyl phosphate. Approximate concentrations of major constitutents in some of the process streams are shown.

this way, nitrite ions are destroyed and plutonium is reduced to the nonextractable Pu^{3+}. The reduction of nitrite ions occurs according to the reaction

$$NO_2^- + NH_2SO_3^- \rightarrow SO_4^{2-} + N_2 + H_2O.$$

The reaction which is responsible for the stripping of plutonium is:

$$Pu^{4+}(organic) + Fe^{2+} \rightarrow Pu^{3+}(aqueous) + Fe^{3+}.$$

The destruction of nitrite ions avoids the two undesired reactions.

$$Fe^{2+} + NO_2^- + 2H^+ \rightarrow Fe^{3+} + NO + H_2O$$
$$Pu^{3+} + NO_2^- + 2H^+ \rightarrow Pu^{4+} + NO + H_2O.$$

During the past few years hydroxylamine has been preferred to ferrous sulpha-mate as stripping agent since iron adds to the waste volume and sulphate ions complicate the plutonium recycle by forming complexes. Sulphate ions also create corrosion problems during the concentration and solidification of the wastes. The major disadvantage of using hydroxylammonium nitrate is the slower reduction kinetics. The redox reaction between hydroxylammonium nitrate and plutonium is:

$$2NH_3OH^+ + 4Pu^{4+}(organic) \rightarrow 4Pu^{3+}(aqueous) + N_2O + H_2O + 6H^+.$$

Uranium is stripped from the loaded organic solvent with $0.01\,M$ HNO_3. In this way 99.99% of uranium is removed from the organic phase. The organic TBP solution is then washed with a basic solution to remove the degradation products of TBP. The aqueous uranium solution is concentrated to $1.8\,M$ and made $1\,M$ in HNO_3 before being fed to the second uranium extraction–decontamination cycle. This cycle provides additional purification from Pu and fission products. Residual quantities of Zr and Nb are removed as well. The final purification of U is obtained by using silica gel or similar adsorbants. The uranium product is converted into oxide by thermal decomposition, either by directly heating the solution or after the precipitation of uranium as ammoniun diuranate with NH_3.

The second plutonium extraction–decontamination cycle similarly provides further purification and decontamination of Pu from U and fission products.

In order to make plutonium once again extractable by TBP, Fe^{2+} and sulphamate ions are destroyed by adding HNO_3 and sodium nitrite. In this way Pu^{3+} is oxidized to the extractable Pu^{4+}. Plutonium is stripped from TBP using hydroxylammonium nitrate, purified from residual fission products by silica gel and concentrated by evaporation or ion-exchange. Finally plutonium is precipitated as oxalate and calcined to plutonium oxide.

12.6 SOLVENT-EXTRACTION PROCESSES FOR THE RECOVERY OF URANIUM FROM LEACH LIQUORS [7–8]

Solvent extraction is used in the processing of uranium ores to concentrate the uranium from its minerals and at the same time to purify it from impurities. The nature of the extractant depends on the type of lixiviant used to leach the ore. In turn the lixiviant is chosen according to the composition of the uranium ore. Presently most uranium ores are leached by sulphuric acid in presence of an oxidizing agent, and this leads to uranium being present in the leach solution as uranyl ion, UO_2^{2+}. It follows that the preferred extractants are long chain tertiary amines or a mixture of di(2-ethylhexyl)phosphoric acid, HDEHP, and TBP. Both extracting systems have the capability of extracting uranium in its highest oxidation state i.e. UO_2^{2+}. The solvent extraction processes utilizing these two types of extractants are called AMEX (amine extraction) and DAPEX (di-alkylphosphoric acid extraction).

12.6.1 The AMEX process [7]

In this process the extractant is a trioctyl- or a tridodecylamine dissolved in an aliphatic diluent such as kerosene. Since the solubility of amine salts in aliphatic diluents is relatively low, a phase modifier, such as decanol or nonanol, is added to the organic solution to prevent third-phase formation. A typical extractant composition is 5% of tridodecylamine (Alamine 336) and 2% of isodecanol in kerosene. A flowsheet for the AMEX process is shown in Fig. 12.7. The leach solution is first

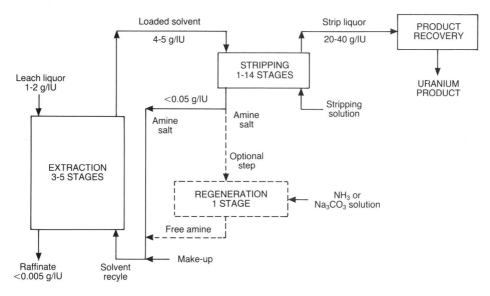

Fig. 12.7 — The AMEX process for uranium extraction. Reprinted from R. Burkina, ed., *Topics of Non-Ferrous Extractive Metallurgy*, Blackwell Scientific Publications, Oxford, 1980, by permission of the copyright holders, the Society of Chemical Industry.

filtered through a sand filter in order to remove most of the suspended solids. If the concentration of suspended solids in the solution contacted with the organic phase is more than 20 ppm, interfacial solid material can be formed, and this would interfere with the proper operation of the extraction equipment.

The chemical reactions which are responsible for the extraction of uranium are:

$$2\,\overline{R_3N} + H_2SO_4 \rightleftharpoons \overline{(R_3NH)_2SO_4}$$

$$\overline{(R_3NH)_2SO_4} + H_2SO_4 \rightleftharpoons 2\,\overline{(R_3NH)HSO_4}$$

$$\overline{2(R_3NH)_2SO_4} + UO_2(SO_4)_3^{4-} \rightleftharpoons \overline{(R_3NH)_4UO_2(SO_4)_3} + 2SO_4^{2-}$$

The bars indicate organic-phase species. The species which most interfere with uranium extraction are the anions of V(V) and Mo(VI). These interfering species are eliminated by reducing V(V) to the nonextractable V^{4+} and by precipitating Mo during the feed pretreatment. The interference from NO_3^- and Cl^- is less important. Uranium can be stripped from the organic phase with an acidic, a neutral or an alkaline stripping solution. When a chloride acidic solution is used ($1M$ NaCl + $0.05\,M$ H$_2$SO$_4$) the stripping reaction is a liquid ion-exchange process

$$\overline{(R_3NH)_4UO_2(SO_4)_3} + 4HCl \rightleftharpoons 4\overline{R_3NHCl} + UO_2SO_4 + 2H_2SO_4.$$

In this case the chloride salt of the amine can be directly recycled to the extraction section. If the stripping solution is $1M$ $NH_4NO_3 + 0.1M$ HNO_3, a solvent regeneration step is required instead.

The neutral stripping solution consists of a concentrated solution of $(NH_4)_2SO_4$. Uranium is brought back into the aqueous phase by the addition of NH_3. In this way the amine is regenerated and the precipitation of uranium is prevented by the high SO_4^{2-} concentration.

Alkaline stripping solutions consist of $0.75M$ Na_2CO_3. In this case the regeneration of the amine and the back extraction of uranium into the aqueous phase occur simultaneously. In the aqueous solution soluble uranium–carbonate complexes are formed.

Uranium can be recovered from any of the stripping solutions by precipitation as ammonium diuranate, $(NH_4)_2U_2O_7$, by bubbling gaseous NH_3.

12.6.2   The DAPEX process [8]

In this process the extractant is a 4% solution of HDEHP in kerosene containing also 4% of TBP. TBP prevents third phase formation during the stripping of uranium, and also gives a synergistic effect with HDEHP in the extraction of UO_2^{2+}. Without TBP, the sodium salt of HDEHP, formed on contact with Na_2CO_3 during the stripping operation, would precipitate, and the distribution ratio of UO_2^{2+} in the extraction step would be too low to be exploited practically. In this process the need for removal of suspended solids is less stringent; up to 100 ppm of suspended solids can be tolerated. The major disadvantage of the process is that HDEHP is less selective for UO_2^{2+} than the tertiary amine. V(IV), Mo(VI), rare earths, Fe^{3+}, Al^{3+}, Ti^{4+} and Th^{4+} are all strongly extracted. Iron(III) is removed by reducing the feed with scrap iron. The precipitate formed is then removed by clarifying the feed by filtration. The flowsheet of the DAPEX process is shown in Fig. 12.8.

The extraction reaction is:

$$UO_2^{2+} + 2\,\overline{(HDEHP)_2} \rightleftharpoons \overline{(DEHP)_2UO_2(HDEHP)_2} + 2H^+.$$

The extraction of UO_2^{2+} is further enhanced by the solvation of the UO_2–HDEHP complex by molecules of TBP. Uranium is stripped from the organic phase by use of a 15% aqueous solution of Na_2CO_3. The stripping reaction is:

$$\overline{(DEHP)_2UO_2(HDEHP)_2} + 2Na_2CO_3 \rightleftharpoons 4\,\overline{NaDEHP} + UO_2(CO_3)_2^{2-}.$$

During the stripping, any iron or titanium present precipitates so the strip solution must be filtered. The UO_2^{2+} ion is kept in solution as the strong tricarbonato complex. The extractant is recycled as its sodium salt. The formation of the sodium salt causes a decrease of acidity in the extraction stages. This decrease must be compensated by adding more acid. Uranium can be finally recovered by first neutralizing the solution with NH_3 and then precipitating it as uranyl peroxide. Uranyl peroxide is then dried to $UO_4 \cdot 2H_2O$ (yellow cake).

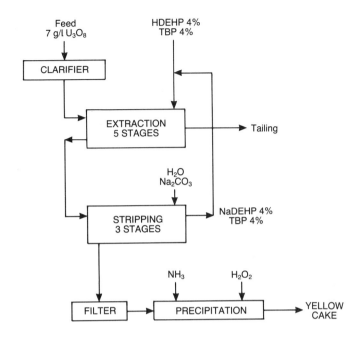

Fig. 12.8 — The DAPEX process.

12.7 SOLVENT EXTRACTION OF ACTINIDES AND RARE-EARTH FISSION PRODUCTS FROM PUREX LIQUID WASTES [9–16]

During the reprocessing of irradiated nuclear fuels by the PUREX process, acid streams containing small amounts of actinides, fission products and corrosion products are generated. Removing and concentrating the long living and radiotoxic transuranium elements, particularly Am, which are present in the high-level liquid waste (HLLW) can make the disposal of the waste much more attractive economically, particularly when ^{90}Sr and ^{137}Cs are also removed. The removal of the actinides allows the waste to be disposed of without the need to bury it deeply below ground, once it has been converted into a solid.

Recently a new class of bifunctional extractants of the carbamoylmethylphosphoryl type has been studied by Horwitz and co-workers [9–15] as potential solvent extraction reagents for the removal of actinides and rare-earth fission products from HLLW. A conceptual solvent-extraction process has also been devised and tested on the laboratory scale [16]. The process is called TRUEX (TRansUranium elements EXtraction).

Some of the new reagents of the carbamoylmethylphosphoryl family are shown in Table 12.8. These reagents look particularly attractive for the treatment of acidic wastes, since they have the ability to extract actinides from rather concentrated nitric acid solutions. Carbamoylmethylphosphoryl extractants can be of the phosphonate, phosphinate and phosphine oxide type, according to the number of alkoxy groups

Table 12.8 — Abbreviations, structures and nomenclature of CMP and CMPO extractants

Abbreviations	Extractant	Nomenclature
DHDECMP	$(C_6H_{13}O)_2P-CH_2-C-N(C_2H_5)_2$ (both P=O and C=O)	Dihexyl-N,N-diethylcarbamoylmethylphosphonate
HHDECMP	$\begin{array}{c}C_6H_{13}\\C_6H_{13}O\end{array}\!\!-P-CH_2-C-N(C_2H_5)_2$ (P=O, C=O)	Hexylhexyl-N,N-diethylcarbamoylmethylphosphinate
DHDECMPO	$(C_6H_{13})_2P-CH_2-C-N(C_2H_5)_2$ (P=O, C=O)	Dihexyl-N,N-diethylcarbamoylmethylphosphine oxide
HΦDECMPO	$\begin{array}{c}C_6H_{13}\\\Phi\end{array}\!\!-P-CH_2-C-N(C_2H_5)_2$ (P=O, C=O)	Hexyl(phenyl)-N,N-diethylcarbamoylmethylphosphine oxide
6-MHΦDECMPO	$\begin{array}{c}(CH_3)_2CH(CH_2)_5\\\Phi\end{array}\!\!-P-CH_2-C-N(C_2H_5)_2$ (P=O, C=O)	6-Methylheptyl(phenyl)-N,N-diethylcarbamoylmethylphosphine oxide
OΦD[IB]CMPO	$\begin{array}{c}C_8H_{17}\\\Phi\end{array}\!\!-P-CH_2-C-N[CH_2-CH(CH_3)_2]_2$ (P=O, C=O)	n-Octyl(phenyl)-N,N-di-isobutylcarbamoylmethylphosphine oxide
6-MHΦD[IB]CMPO	$\begin{array}{c}(CH_3)_2CH(CH_2)_5\\\Phi\end{array}\!\!-P-CH_2-C-N[CH_2CH(CH_3)_2]_2$ (P=O, C=O)	6-Methylheptyl(phenyl)-N,N-di-isobutylcarbamoylmethylphosphine oxide
2-EHΦD[IB]CMPO	$\begin{array}{c}C_4H_9CH(C_2H_5)CH_2\\\Phi\end{array}\!\!-P-CH_2-C-N[CH_2CH(CH_3)_2]_2$ (P=O, C=O)	2-Ethylhexyl(phenyl)-N,N-di-isobutylcarbamoylmethylphospine oxide

that are attached to the phosphorus. The chemical stability and the basicity of these extractants are in the order phosphine oxide > phosphinate > phosphonate. Their ability to retain a high extraction power in presence of high aqueous acid concentrations is attributed to the internal buffer capacity of the group CO–N which, in presence of a strong acid, can be protonated to yield CO–$\overset{+}{\text{N}}$H. In this way the P=O group remains at least partially available for actinide and lanthanide extraction since, differently from TBP, the extracted HNO_3 mainly interacts with the amide part of the molecule. The reagents which show the best extraction performance are the carbamoylmethylphosphine oxides. In particular octyl(phenyl)-N,N-di-isobutyl-carbamoylmethylphosphine oxide (OØD[IB]CMPO or just CMPO) demonstrates the best extractive performance. The particular nature of the substituents present on the molecule make this compound very stable to hydrolytic and radiolytic degradation. In addition, CMPO has good selectivity for actinides over many fission products and favourable solubility and loading properties. The good performance of CMPO is due to the optimal basicity of the phosphoryl group, which is partially reduced by the inductive effect of the benzene ring. To improve the solubility of CMPO in the aliphatic diluents which are used in the TRUEX process, to reduce the possibility of third-phase formation, and to increase the loading capacity of the organic phase, the reagent is used in presence of TBP which acts as a phase modifier. The amount of modifier can be optimized according to the acidity of the feed and the nature of the diluent. The distribution coefficient of Am(III) by phosphonate, phosphinate and phosphine oxide reagents in mixture with TBP as a function of the aqueous HNO_3 concentration is shown in Fig. 12.9. Although the presence of TBP reduces to some extent the extractive power of the reagent, CMPO still retains good extractive properties, owing to its higher basicity. The influence of TBP on the extraction of Am by CMPO is shown in Fig. 12.10. When the concentration of TBP is $0.75\,M$ and that of CMPO is $0.25\,M$ the organic phase has optimal extractive behaviour in view of the reduced influence that the aqueous HNO_3 concentration has on the extraction of Am in the acid range 0.5–$6\,M$. This improved performance in presence of TBP is due to the reaction of HNO_3 with TBP. The chemical reactions describing the extractive behaviour of CMPO in presence of TBP from nitric acid are (for tervalent cations):

$$M^{3+} + 3NO_3^- + 3\overline{E(HNO_3)_m} \rightleftharpoons \overline{M(NO_3)_3E_3(HNO_3)_n} + (3m-n)\,HNO_3$$

$$HNO_3 + \overline{TBP} \rightleftharpoons \overline{TBPHNO_3}$$

where E = CMPO and the bar indicates organic species.

Since the mixture of TBP and CMPO shows good selectivity for transuranium elements and rare-earth fission products it can be used in the TRUEX process for the removal of these elements from HLLW. A typical composition of an HLLW is shown in Tables 12.9 and 12.10. The distribution ratios (measured by Horwitz *et al.* [16]) for most of the fission products present in HLLW, for the corrosion products Cr, Fe, Ni, and the actinides, between $0.25\,M$ CMPO + $0.75\,M$ TBP in decalin and $1.5\,M$ HNO_3 are reported in Table 12.11. The data have been obtained at 50°C, which is approximately the temperature of HLLW. (This temperature is the result of the heat generated by the radioactivity of the radionuclides.) The distribution ratios in

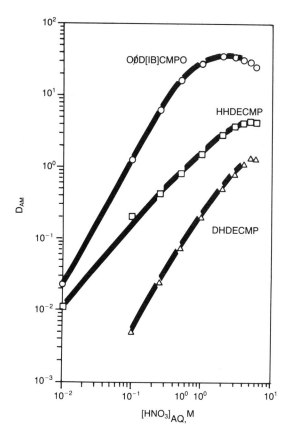

Fig. 12.9 — Acid dependence for Am^{3+} extraction by $0.20M$ CMPO or CMP/$0.8M$ TBP in dodecane, 25°C. Reprinted from E. P. Horwitz and D. G. Kalina, *Solvent Extraction and Ion Exchange*, 1984, **2**, 179, by permission of the copyright holders, Marcel Dekker Inc.

presence of oxalic acid are also reported. Oxalic acid is added to the aqueous phases since it forms complexes with Fe, Zr and Mo, thus lowering their extractability. The flowsheet of the TRUEX process developed by Horwitz is shown in Fig. 12.11. The flowsheet operates at high acidities where the extraction of Tc, Ru and Pd is very low. Fe, Zr and Mo are stripped from the organic phase with oxalic acid and HNO_3. The removal of Fe, Zr and Mo from the raffinate can be very helpful when Ru, Rh, Pd and ^{90}Sr and ^{137}Cs are to be recovered. Plutonium and the transplutonium elements are removed by using a solution containing formic acid and hydroxylammonium nitrate, HAN, $0.05M$. HAN is used to reduce Pu(IV) to the less extractable Pu(III). The process can remove 99.99% of the tervalent actinides. The removal of Np and Pu is sufficiently good to reduce the activity of the raffinate, once transformed into

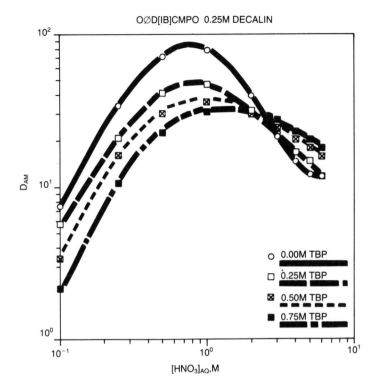

Fig. 12.10 — The influence of TBP on the extraction of americium by CMPO. Reprinted from E. P. Horwitz and D. G. Kalina, *Solvent Extraction and Ion Exchange*, 1984, **2**, 179, by permission of the copyright holders, Marcel Dekker Inc.

Table 12.9 — Molar concentrations of ions in an acidic nuclear waste solution

H^+	1.0
Al^{3+}	0.78
Fe^{3+}	0.16
Cr^{3+}	0.015
Ni^{2+}	0.007
Na^+	0.21
SO_4^{2-}	0.33
F^-	0.14
$NO_3^- + NO_2^-$ } Balance of anions }	~ 3
Actinides (U, Pu, Am, Cm)	~ 5×10^{-4}

Table 12.10 — Molar concentrations of fission products in an acidic nuclear waste solution

Ge	3×10^{-7}	In	0.5×10^{-6}
As	0.5×10^{-7}	Sn	2×10^{-5}
Se	0.25×10^{-4}	Sb	0.45×10^{-5}
Br	0.8×10^{-5}	I	1×10^{-4}
Rb	2×10^{-4}	Cs	0.8×10^{-3}
Sr	0.4×10^{-3}	Ba	0.6×10^{-3}
Y	2.5×10^{-4}	La	0.4×10^{-3}
Zn	1.8×10^{-3}	Ce	0.8×10^{-3}
Mo	1.6×10^{-3}	Pr	0.36×10^{-3}
Te	0.35×10^{-3}	Nd	1.25×10^{-3}
Ru	1×10^{-3}	Pm	0.85×10^{-5}
Rh	1.7×10^{-4}	Sm	0.25×10^{-3}
Pd	0.8×10^{-3}	Eu	0.5×10^{-4}
Ag	2.5×10^{-5}	Gd	0.35×10^{-4}
Cd	0.3×10^{-4}	Tb	0.55×10^{-6}

Table 12.11 — Distribution ratios of selected elements in synthetic PUREX–HLLW at 50°C [16]

	Feed (HLLW) O/A = 0.5	$1.5M$ HNO$_3$–0.75M H$_2$C$_2$O$_4$ O/A = 2
Sr	<0.01	
Y	1.7	1.1
Zr	15	<0.01
Mo	10	<0.1
Tc	0.30	
Ru	0.05	
Rh	<0.02	
Pd	<0.02	
Ag	<0.1	
Cd	<0.01	
Ba	<0.01	
Cr	<0.05	
Fe	5	<0.01
Ni	<0.1	
La	0.72	2.2
Ce	1.4	3.4
Pr	1.9	3.8
Nd	2.4	3.9
Sm	3.1	3.9
Eu	3.4	3.7
Gd	2	
U	$> 10^2$	$> 10^2$
Np	62	13
Pu	$> 10^3$	39
Am	3.9	6.6

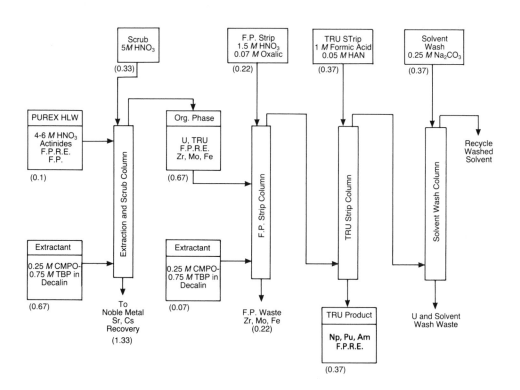

Fig. 12.11 — Flowsheet for the recovery of TRUs (Np, Pu, Am) and fission product R.E.s from PUREX high-level waste [16].

grout, to 100 nanocuries/gram. The uranium which is still present in the organic phase is finally removed with Na_2CO_3, during the solvent wash step.

Acknowledgement.   The author thanks E. P. Horwitz for reviewing the manuscript.

SELECTED REFERENCES

Irradiated nuclear fuel
[1] G. R. Choppin and J. Rydberg, *Nuclear Chemistry*, Pergamon Press, New York, 1980.
[2] S. Peterson and R. G. Wymer, *Chemistry in Nuclear Technology*, Addison-Wesley, Reading, Massachusetts, 1983.

PUREX process
[3] A. Naylor and D. D. Wilson, *Recovery of Uranium and Plutonium from Irradiated Nuclear Fuel*, in *Handbook of Solvent Extraction*, T. C. Lo, M. H. I. Baird and C. Hanson, eds, John Wiley and Sons, New York, 1983, pp. 783–798.

[4] J. M. Cleveland, *The Chemistry of Plutonium*, American Nuclear Society, 1979.
[5] J. M. McKibben, *Chemistry of the PUREX process*, Am. Chem. Soc. 135th National Meeting, March 20–25, 1983, Seattle, Washington.
[6] T. H. Siddal III, in *Chemical Processing of Reactor Fuels*, J. F. Flagg, ed., Academic Press, New York, 1961, Chapter V.

Processes for the recovery of uranium from leach liquors
[7] A. R. Burkin, *Extractive Metallurgy of Uranium*, in *Topics of Non-Ferrous Extractive Metallurgy*, R. Burkin, ed., Blackwell Scientific Publications, Oxford, 1980.
[8] C. A. Blake, C. F. Baes, Jr., and K. B. Brown, *Ind. Eng, Chem.*, 1958, **50**(12), 1763.

Extraction of actinides and rare earths by carbamoylmethylphosphoryl-type extractants
[9] E. P. Horwitz, H. Diamond and D. Kalina, *Plutonium Chemistry*, ACS, Washington, D.C., 1982
[10] E. P. Horwitz, D. G. Kalina, L. Kaplan, G. W. Mason and H. Diamond, *Sep. Sci. Technol.*, 1982, **17**, 1261.
[11] D. G. Kalina, E. P. Horwitz, L. Kaplan and A. C. Muscatello, *Sep. Sci. Technol.*, 1981, **16**, 1127.
[12] E. P. Horwitz and D. Kalina, *Solv. Extr. Ion Ex.*, 1984, **2**, 179.
[13] W. W. Schulz and J. D. Navratil in *Recent Developments in Separation Science*, Vol. VII, N. N. Li, ed., CRC Press, Boca Raton, Florida, 1981; p. 31.
[14] R. R. Shown, W. J. McDowell and B. Weaver, *International Solvent Extraction Conference, Proceedings*, Toronto, Canada, 1977. Vol. 1, p. 101.
[15] W. W. Schulz and L. D. McIsaac, *International Solvent Extraction Conference, Proceedings*, Toronto, Canada, 1977. Vol. 2, p. 619.
[16] E. P. Howitz, H. Diamond, D. Kalina, L. Kaplan and G. W. Mason, *International Solvent Extraction Conference, Proceedings*, Denver, Colorado, 1983; p. 451.

Index

Index